DATE DUE

OC 17 '96			
NO 21 '96			
MR 24 '97			
JE 7 '01			

DEMCO 38-296

Our Universes

OUR UNIVERSES

Sir Denys Wilkinson

Columbia University Press   New York

Columbia University Press
New York   Oxford

Library of Congress Cataloging-in-Publication Data
Wilkinson, Denys Haigh, Sir.
Our universes / Sir Denys Wilkinson.
p.   cm. — (The George B. Pegram lecture series)
Lectures presented at Brookhaven National Laboratory, Feb. 1989.
Includes index.
ISBN 0-231-07184-1
1. Physics—Philosophy.   2. Science—Philosophy.   3. Man.
4. Cosmology.   I. Title.   II. Series.
QC6.W5766   1991
530—dc20   91-35
                                                   CIP

Casebound editions of Columbia University Press books are Smyth-
sewn and printed on permanent and durable acid-free paper
∞
Printed in the United States of America
c 10 9 8 7 6 5 4 3 2 1

The George B. Pegram Lecture Series

To provide an opportunity for distinguished scholars to examine the interaction between science and other aspects of our culture and society, the Trustees of Associated Universities, Inc., established the George B. Pegram Lecture Series at Brookhaven National Laboratory.

The lectures are named in honor of George Braxton Pegram, who contributed so much to our country in general, and to Brookhaven National Laboratory in particular. Except for a few years abroad, George Pegram's entire professional career was spent at Columbia University, where he was a Professor of Physics, Dean and Vice President. In 1946, he headed the Initiatory University Group, which proposed that a regional center for research in the nuclear sciences be established in the New York area. Thus, he played a key role in the founding of Brookhaven and became one of the incorporating trustees of Associated Universities, Inc., remaining an active trustee for ten years.

George Pegram devoted his life to physics, teaching and the conviction that the results of science can be made to serve the needs and hopes of humankind.

George B. Pegram Lecturers

1949	Lee Alvin DuBridge	1969	Roger Revelle
1960	René Jules Dubos	1970	Barbara W. Tuchman
1961	Charles Alfred Coulson	1971	George E. Reedy
1962	Derek J. deSolla Prince	1972	Colin Low
1963	J. Robert Oppenheimer	1975	Jean Mayer
1964	Barbara Ward	1979	Sir Peter Medawar
1965	Richard Hofstadter	1985	David Baltimore
1966	Louis S. B. Leakey	1988	Robert C. Gallo
1968	André Maurois	1989	Sir Denys Wilkinson

Contents

Preface

We look outward to the Universe and inward to ourselves. The Pegram Lectures, given at Brookhaven National Laboratory in February 1989, were an exploration of the relationship between these two activities; they cannot be separated. Immanuel Kant's dichotomy of "starry heavens above" and "moral law within" are but the two sides of the same coin.

Just as we cannot meaningfully define a physical law in the absence of matter to which that law applies, so we cannot meaningfully define a Universe in the absence of a mind to make that definition. But minds are finite and must be limited in the range of concepts of which they are capable so that the conclusions that we can reach about the nature of the Universe must, in qualitative terms, be limited by the nature of our minds. In this sense we do not access an objective external Universe but rather construct it from the materials of our own nature: the Universe as artifact. Nature outside us is constructed from Nature inside us and in effecting that construction our tastes and feelings must enter just as surely and ultimately involve the same essential internal criteria as they do in judgments of an artistic or personal kind.

But the Universes of our mind are not limited to that in which that mind resides. In recent years physics has led us inexorably toward the concept of multiple Universes to the degree that today the idea that we reside in a Universe that is unique other than in being the seat of our abode is becoming as self-centered a concept as the cosmology of Ptolemy. We cannot choose in which Universe we wish to live, but the conditions that must be satisfied by a Universe within which we might have emerged are so many and so tight that it is true to say that the Universe chooses us.

This book is an expanded version of the lectures, but I have chosen to keep it in narrative style rather than give it any pedogogical content; I have similarly chosen to eschew the scholarly machinery of references, footnotes, and so on in an attempt to make its message the more immediate: that we are incomprehensibly privileged to be here in a cosmos for which we must accept responsibility.

I have also chosen to use a large number of quotations. This is not so much to recognize T. S. Eliot's dictum:

One can always save the subject by magnificent quotations.

as to illuminate and help draw together the rather widespread strands of my argument through reference to familiar antecedents, often in different contexts; the quotations are not pegs upon which the argument is hung but rather marker posts that I leave behind as reminders of the path that I have taken. But I also use quotations when I feel that it is best to express the thought through the *ipsissima verba* of the original voice. There are, therefore, few quotations in the more straightforwardly expository sections of the book, whose primary purpose is to establish the physical context for my main themes of the structure of our judgments about the nature of the physical world and the conjunction between the physics of our local context and the metaphysics of the infinitude of our Universes.

While I find it is not possible for me to identify, with any confidence, all the influences that must, over the years, have contributed to the thoughts I present in this book, I should like to say that in my recent reading that has led to the actual writing I have particularly benefited from: John Barrow and Frank Tipler's *The Anthropic Cosmological Principle* (1986); Jerome Bruner's *Actual Minds, Possible Worlds* (1986); and Richard Gregory's *Mind in Science* (1981).

Our Universes

1. Our Nature

Doubt

In 1628 John Donne wrote:

The difference between the reason of Man and the instinct of the beast is this, that the beast does but know but Man knows that he knows.

Three centuries later this sentiment was echoed by Teilhard de Chardin:

An animal knows, of course. But certainly it does not know that it knows.

A few years after Donne, René Descartes put it more succinctly:

Cogito ergo sum.

"I think therefore I am"; that is to say our consciousness of ourselves implies our own existence and therefore that of the Universe, the context within which that consciousness is itself defined. Today, as we question more deeply the nature of existence, both our own and that of our setting, the material Universe, as we wonder whether our consciousness and sentience are not just, literally, the output of a computer program, as we wonder whether the Universe in the gross, the whole shebang, is not just the output of a computer program, doubt creeps in. We recall that Goethe remarked:

Doubt grows with knowledge.

and that Blaise Pascal concurred with:

The last function of reason is to recognize that there are an infinity of things which surpass it.

with Miguel de Unamuno capping it by:

The supreme triumph of reason . . . is to cast doubt upon its own validity.

I say to myself:

I used to think that I was indecisive, but now I'm not so sure.

and "I think, therefore I am" is replaced by:

I think, therefore I am, I think.

Perhaps I should leave it with Paul Valéry:

Sometimes I think: and sometimes I am.

Nature

My purpose in this book while, I hope, retaining my card as a professional physical scientist, is to look into the relationship between the physical Universe, if there is such a thing, and our perception of it, to relax the usual tacit assumption of the physical scientist that there is indeed an objective physical Universe "out there," structured and functioning independently of our perception of it, that we may access and describe with an objectivity that is independent of our own nature so that an identical account would be given by sufficiently intelligent woodchucks or AIDS viruses.

I exaggerate: one can confidently, using information theory, relate the essential complexity of a system to be described to the necessary complexity of the system effecting the description and this probably eliminates AIDS viruses as serious contenders for cosmic Nobel prizes. It does not, however, eliminate woodchucks.

And why, by extrapolation from woodchucks, which we may regard as lovable but not too bright, through human beings, whom we regard as significantly brighter, should we assume that we, bright as we think ourselves to be, represent an evolutionary development adequate for gaining total comprehension of the world outside such that our description of it may then be said to be objective and complete? Rather should we remember what Charles Darwin said:

... can the mind of Man ... developed from a mind as low as that possessed by the lowest animal, be trusted when it draws such grand conclusions?

We can look at the Universe only through our own eyes, and so all that we can know about the Universe is what we, literally or metaphorically, see through our own eyes. The only objectivity that the Universe displays is what it displays to us. All that we can make of the Universe is what we make of it as human beings. We must, therefore, ask how we think, how we know and what might be the limits imposed by our nature on that thought and knowledge.

I shall echo Alexander Pope in the first lines of his *Essay on Man:*

Know then thyself, presume not God to scan;
The proper study of mankind is man.

And I shall echo him in the last line of that same *Essay:*

And all our knowledge is, ourselves to know.

As James Clerk Maxwell asked in his inaugural lecture as Cavendish Professor in Cambridge in 1871:

... is the student of science to be withdrawn from the study of man?

Ourselves as part of Nature and Nature as part of ourselves is the burden and leitmotiv of this book. I shall emphasize it repeatedly and tediously, but not sufficiently, in my attempt to give the lie to Henry David Thoreau when he wrote:

It appears to be a law that you cannot have a deep sympathy with both Man and Nature.

What I shall be saying is that in our attempt to reach out to the Universe, that very sympathy is not only possible but also essential; human beings and Nature cannot be untwined.

Bewitched by Nature

One of my favorite verses is by the Russian poet N. Olennikov:

> Not for you are passion and goldlust
> It is science that entices you.
> Passion may fade and love is betrayed
> But you cannot be deceived
> By the bewitching structure of the cockroach.

Although, in this book, I shall not explicitly address problems to do with passion and goldlust I shall attempt to convince you that we may well *be* being deceived by the bewitching structure of the cockroach; Nature attempts to seduce us by those same voices that the far-off heroes heard and may lead us, because of our human frailty, into false conclusions about Nature. But then, in revenge, it is that same human frailty that, inescapably, determines the Nature that we construct. Recall Hilaire Belloc's little verse:

> The Devil, having nothing else to do,
> Went off to tempt my Lady Poltagrue.
> My lady, tempted by a private whim,
> To his extreme annoyance, tempted him.

My Texts

Back to the ostensibly more serious. If I might, additionally, take a text for this book it would be from Albert Einstein:

The fairest thing that we can experience is the mysterious. It is the fundamental emotion which stands at the cradle of true art and true science.

To which I feel that I should append, as a kind of trailer of what is to come, the aphorism of Michel Eyquem de Montaigne, pronounced, consistently with Einstein, four centuries earlier but even more true today than then:

Wonder is the foundation of all philosophy, inquiry the progress, ignorance the end.

Or, as Alfred North Whitehead put it a little less pessimistically:

Philosophy begins in wonder. And, at the end, when philosophic thought has done its best, the wonder remains.

Idealism

Physicists practice scientific method by definition and by the light of Nature, more in ignorance of it than by precept and not benefiting much from the efforts of others at its codification. Nor are physicists usually much concerned with the nature and mechanisms of their own thought; with perception and cognitive processes. And this, by and large, is as it should be because it does not help one's work to realize how limited are the tools available for it. But that is just what I aim to explore in this book: to ask to what degree the doing of the job may be limited by the tools and the nature of the worker.

We must indeed recognize that we cannot stand back: we are part of the system that we are describing and what we say about that system, that Universe, cannot be uncoupled from our perception of it, from our own nature. This does not mean that one must subscribe to any of the philosophical doctrines of idealism that are, in their simplest form, associated with Bishop Berkeley, the eighteenth-century philosopher-divine who held that material things do not exist independently of minds to observe them: finite human minds if such are around, failing which the infinite mind of God takes over, as succinctly expressed by Ronald Knox from his Oxford cloister:

There once was a man who said "God
Must think it exceedingly odd
    If he finds that this tree
    Continues to be
When there's no one about in the Quad."

and its anonymous riposte:

"Dear Sir, Your astonishment's odd
I am always about in the Quad.
    And that's why the tree
    Will continue to be
Since observed by, Yours faithfully, God."

I stray into naive idealism to emphasize how essential it is to recognize that our only contact with the Universe that we seek to understand must come, directly or indirectly, through our own perceptions of it, that those perceptions are our only evidence that there is something out there, that, conversely, in the absence of those perceptions we have, by definition, no evidence that there is anything there and therefore no reason to suppose that it continues to exist; no reason of course, except common sense.

For my purposes it is salutary to pause at this extreme point of naive idealism, which is absurd but which, as Boswell observed before Dr. Johnson hurt his toe, we cannot refute; it concentrates the mind on the essential nature of the human being/Universe relationship that I wish to explore; we must abandon that absurd viewpoint of naive idealism only cautiously and consciously, aware that, as Henri Gaudier-Brzeska wrote from the trenches of World War I with something different in mind but to the same effect:

Much will be changed when we have come through the blood-bath of idealism.

Our Isolation

We must also recognize the other irrefutable lesson of idealism, namely, the essential isolation of the individual observer, the individual human being. In its logically extreme form this means me but not you; you are part of my external world; I can gain no direct evidence of your consciousness and do not need to concern myself with it any more than I concern myself with the existence, or otherwise, of the chair upon which I now sit to write. You are simply part of my context, my environment. This form of idealism is solipsism: nothing exists but me and my mental states. Much philosophy skirts perilously close to solipsism. But whether or not the existence of you, my reader, and the rest of the Universe depends on my consciousness, it is certainly true that it is only through that consciousness that I can be aware of the existence of the Universe. We cannot look elsewhere for understanding than to our separate selves. As Webster's character remarks in *The White Devil:*

> "I do not look
> Who went before nor who shall follow me
> No, at myself I will begin and end."

It is true that those who went before me aided me in my understanding and that those who follow me may in turn benefit from my understanding, but the individual confronts the world alone and only he or she commands knowledge of it; only he or she can experience it and can experience understanding of it.

The Commanding of Knowledge

How do we command knowledge? How do we think? How do we know that we know? What, literally, goes on in our heads? What are the filters, of whatever kind, neurophysiological, psychological, cultural, that may stand between our receipt of signals from that world outside our individual selves and their processing into our conclusions about their nature, relevance, and meaning? What are the mechanisms through which thoughts are generated, through which externality is internalized and then reexternalized as communication?

If we understand, even only partially, the answers to questions such as these we may also understand some of the limitations that our own structure and human nature place upon the degree to which the Universe might be comprehensible to us and even gain some glimmerings of what it is that we cannot know. For, as Ludwig Wittgenstein famously remarked in the last proposition of his *Tractatus Logico-Philosophicus:*

Man must be silent about that of which he cannot speak.

This proposition, most importantly for us in our addressing of the limits to knowledge, leaves open the possibility that we may recognize that there indeed *do exist* books that must remain forever closed, questions to which there can be no answer except their contemptuous but uneasy dismissal: "What happened before the beginning of time?"

Philosophy

Before I begin my examination of the roots of our understanding of the natural world I should say that I am going to adopt a broadly philosophical position. In saying this I am using "philosophy" in the strict sense of its definition in that highest of authorities, the *Oxford Dictionary of English Etymology,* namely, "the study of things and their causes." Although I do not believe that commitment to specific philosophies has been particularly helpful to the de facto development of science, except occasionally and by accident, I do hold very strongly that the qualitative but rigorous discipline of philosophy, which demands demonstration and questions beliefs, must illuminate all attempts to broaden the horizon of our understanding. What F. Zwicky says is all too often all too sadly true:

Once Man is in a rut he seems to have the urge to dig ever deeper.

The job of philosophy is to get us out of that rut. Hearken to Bertrand Russell:

He who has no tincture of philosophy goes through life imprisoned in the prejudices derived from common sense, from the habitual beliefs of his age or his nation, and from convictions that have grown up in his mind without the cooperation or consent of his deliberate reason. As soon as we begin to philosophize, on the contrary, we find that even the most everyday things lead to problems to which only very incomplete answers can be given. Philosophy, though unable to tell us with certainty what is the true answer to the doubts which it raises, is able to suggest many possibilities which enlarge our thoughts and free them from the tyranny of custom.

In the sense of the etymological dictionary and Bertrand Russell, this book is about philosophy. Even more so it is a book on human beings. In my view, our psyche, and the feeling through which that psyche expresses itself to our consciousness, does not generate philosophy, which then generates science: rather our feeling generates our science and our philosophy independently and in parallel; they later, remaining distinct, interact symbiotically with occasional suggestions of syzygy but not of symphysis. I hope that that clarifies my position.

We should not fear philosophy. Many have found it impossible to separate science and philosophy at the frontiers of knowledge and understanding. Max Born writes:

I am now convinced that theoretical physics is actual philosophy.

and D'Arcy Thompson:

Physical science and philosophy stand side by side, and one upholds the other. Without something of the strength of physics philosophy would be weak; and without something of the philosopher's wealth physical science would be poor.

and he quotes approvingly J. H. Fr. Papillon:

Nothing will remove from the tapestry of science the golden threads put there by the hand of the philosopher.

There is certainly no universally correct philosophy for all forms of scientific endeavor. Rather there is a kind of continuum of philosophies that is filtered by the sorts of problem that are referred to it.

Cognitive Science

What, then, do we know about how we think and feel? This is the domain of cognitive science, one of today's major academic growth industries linking neurophysiology, psychology, philosophy, and computer science. Cognitive science attempts to construct operational accounts of those processes, essential to the development of our understanding of our context in the natural world, that are usually given fuzzy labels such as "intuition" and "creativity" and to which I shall, from time to time, refer. I shall not be so unwise as to attempt a conspectus of cognitive science but shall rather just dip in at one or two points that bear particularly upon my broader task.

A field that so is broad and so well tilled that I will barely mention it concerns imagery. Some people think in images or pictures and some do not, or, more particularly, claim that they do not. This goes back to Aristotle who asserted:

Thought is impossible without an image.

Some have disagreed. There are interesting differences of national style. Thus the Anglo-Saxons are thought of as the great image makers. And it is true that the beginnings of electromagnetism are to be found in the imagination-gripping but wholly unphysical lines of force of Michael Faraday, who said:

Nothing is too wonderful to be true.

However, when these were taken over and wrestled with by James Clerk Maxwell they turned in 1864 into the abstractions of the first of the field theories that today dominate the scene in fundamental science. Today when we similarly wrestle with the problems of the forces between quarks and the mystery of their confinement within the proton and other particles we invoke images of an identical nature to those of Faraday. Conversely, the Latins are thought of as the exploiters of logical argument and method, but what could be more concrete than the images of Leonardo da Vinci or Enrico Fermi? Some of the architects of our most wonderful and abstract intellectual structures have insisted on their dependence on quite homespun pictures and metaphors. Thus Albert Einstein, who once asked, perhaps not entirely innocently

What, precisely, is thinking?

always insisted on the primacy of visual images in his work—although he also said:

. . . the creative principle resides in the mathematics.

Equivalences

Very often the routes of visual imagery and abstraction have come together and been shown to be equivalent despite their apparently radically different formulations: Schrödinger's visualizable atom and Heisenberg's matrix mechanics; Feynman's diagrammatic quantum electrodynamics and Schwinger's Green's functions. This suggests strongly that the input at the level of consciousness does not matter and that what is of greater significance is the mechanism by which the concepts, however expressed, are manipulated at a deeper level following, as we shall now see, some coding process.

The Coding of Mental Operations

A three-and-a-half-year-old girl was asked to copy a drawing of a square. Figure 1 shows the results. She was asked to do it again

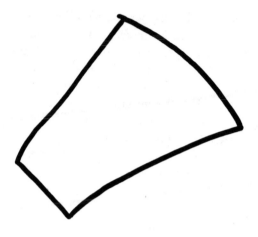

Figure 1

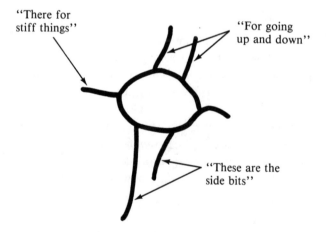

Figure 2
From Jean Hayes, *Cognitive Science* 2 (1978).

and produced figure 2, which also shows her explanation of what her second attempt meant. When asked which, in figure 1, were the "stiff things," the things "for going up and down," and the "side bits" of figure 2, the girl indicated that these were the square's corners, verticals, and horizontals, respectively. The external world, gaining access to our internal processing machinery via our sense organs, is evidently encoded in ways such as this.

We are not ourselves consciously aware of the processes through which sensory inputs are coded, processed, stored, retrieved, re-processed, decoded, and converted into sensory outputs such as happens when we look at and draw a square. Some clues are forthcoming from young children in whom the various processes are still to some degree accessible to consciousness and who have acquired enough language to express that consciousness. Other clues come from victims of brain damage or from persons to whom some sense has been granted late in life so that, for example, sight is suddenly added to touch.

The apparently straightforward effort of figure 1 is the result of an even more complicated effort than figure 2 because it involves an additional stage, namely, reconstruction from the symbolic description. Evidence such as this shows unequivocally that the human brain operates via a symbolic coding mechanism much as computers do. This is why some cognitive scientists reckon that computers should be able, in a fundamental way, to help us with our thinking. This is as yet an unproven assertion, but not yet a forlorn hope.

I am sure that most people feel that they copy a drawing of a square by somehow tracing the image off the retina. It is not at all like that: it is much more complicated. And how very much more complicated again must be the gymnastics within our brains that enable us to generate and process the thoughts relating to the normal activities of our everyday lives let alone multidimensional non-Euclidean geometry and the abstract concepts that lie behind our discussions of the submicroscopic world and the earliest moments of creation to which I shall turn later on. We have no possibility of accessing such inner processes and so, to anticipate slightly one of the major points I am going to make, no possibility of knowing what limitations our mechanisms of ratiocination may not lay upon the sorts of thought we are capable of having.

Our brains have developed coding procedures primarily to enable them to deal with the standard problems of living, but they also serve to deal with contingent matters of a higher order of abstraction such as non-Euclidean geometry, genetic manipulation, and, even more so, the ethics of genetic manipulation—for our brains also somehow handle love and hate. Some of the basic wiring of our brains is doubtless genetically inherited and determined, but evidently we can extend that wiring or develop new chunks of software as experience grows, and recent research has shown that both methods are possible, establishing symbolic coding for sufficiently persistent aspects of life as it reveals itself to us. Our brains are obviously capable of this continuing receptivity, are capable of accommodating and operating upon inputs for which they were not prewired. I am completely confident that my DNA does not speak English nor even, after sixty years' practice, my neurons. We are certainly not born into this world with brains already kitted out with the equations of general relativity, but our brains have the capacity to establish an appropriate symbolic coding so that we can assimilate those equations and, through that coding, give to the equations, to the entities that the equations represent, and to the concepts that are encapsulated in the equations the possibility of development, processing, and eventual outputting at the level of our consciousness so that we can do something about it.

We have no idea if we all code the same way. That we might in the case of drawing squares is not unreasonable because that type of task is one that we must all face so that the mechanism for carrying it out may well be incorporated in the genetic material. It is thought by some psychologists that quite complicated matters are inherent in such a way. Thus C. G. Jung:

In every individual . . there are the great "primordial images," the inherited potentialities of human imagination. They have always been potentially latent in the structure of the brain. The fact of this inheritance also explains the otherwise incredible phenomenon, that the matter and themes of certain legends are met with the whole world over in identical forms.

And we have Noam Chomsky's related view of the Deep Structure of language as a built-in inheritance of the brain. But it seems to me to be quite unreasonable to imagine the same thing about the equations of general relativity, which are probably processed in very different ways in different brains. Some brains are doubtless better constituted for coding certain sorts of tasks and concepts than others: general relativity or painting or playing tennis or the drafting of arms control agreements. That is what makes the difference between us; the trick is finding what our individual brains happen to be best at.

The Brain's Limited Powers

The processes of evolution that brought our brains to their present condition did not have specific regard for general relativity, which brings rather little evolutionary benefit to the species, but it does turn out that our brains can be set up to cope with it. However, our brains, wonderful as they are, are limited structures and so must be capable of only a limited number, and a limited nature, of recodings. It must, therefore, be that there are certain things about which we intrinsically cannot think because our brains are intrinsically incapable of coding the concepts that would be necessary to accommodate those things. It is not that we are not clever enough to think of those things, it is that our brains are simply incapable of the necessary additional programming, not so much because of their physical size as because of their nature. We should be extraordinarily arrogant to think otherwise.

Although those inaccessible things may be of great importance in the domestic economy of the Universe and may give rise to obvious phenomena that we can detect and study, we shall not be able to formulate the necessary concepts to describe those things; for us those things simply do not and cannot exist. I shall describe the phenomena to which they give rise by new theories of a kind permitted to us by the nature of our mental equipment, or by the tacking of supplementary hypotheses onto our existing theories. But that is all I shall be able to do. You will forgive me if I do not give examples of those inaccessible things.

It may be that the process goes only part way. In the case of the child drawing the square, the symbolic coding takes place—the child has "taken in" the square—then there is the decoding and reconstruction leading to consciousness and motor action—the square is "outputted." Perhaps in the case of some of the unspeakable things of my present argument we can code, we can take in, but then we cannot decode and raise to the level of consciousness: we do not know that we know.

Such a possibility of not knowing that we know has recently been demonstrated in the context of the condition known as prosopagnosia—the inability of some people to recognize familiar faces. The recent discovery is that although such persons cannot recognize the face at the level of their conscious awareness, certain autonomic changes, such as a change in the conductivity of their skin, on being shown the picture of the face show that the face has indeed been recognized; the brain has responded but has been unable to communicate that response to the consciousness: the person does not know that he or she knows.

So also with my unspeakable thoughts: we may be having the thoughts but not know that we are having them. And if those thoughts in fact were to break through into our consciousness how should we recognize them as acceptable additions to the rationalization of our experience and not to be dismissed as fantasy? Goethe writes:

What Man does not know,
Or has not thought of,
Wanders in the night
Through the labyrinth of mind.

In the most famous of Goya's *Caprichos* a slumped figure is surrounded by terrifying flying creatures; the picture is entitled "The sleep of reason brings forth monsters." Perhaps it is like that with unthinkable thoughts. Perhaps woodchucks could do better: how can we tell? Perhaps such thoughts are indeed possible for a few people whose brains have developed very unusual codings: we call them madmen and madwomen.

If one wishes to look further ahead into the development of the human race than I do here one might speculate about the possibility of picking up unspeakable thoughts by autonomic changes to which they give rise when they cannot reach the level of consciousness; such changes would naturally be best registered by computer, so we might conceive of a new dimension of human thought in which consciousness is replaced by a computer output, whose reading would, in a certain sense, then raise the unthinkable thought to the level of consciousness at one remove.

Unrecognizable Prejudice

We must also recognize the other side of this coin. Just as there are certain new wirings, new software, and new codings that our brains cannot accommodate, so there may be old wirings, old codings, perhaps genetically transmitted or acquired in the universal cradle, that cannot be replaced or overridden. These will be important codings, for example, to cope with the exigencies of everyday existence: food, defense, reproduction, and so on, and to represent our common sense. But these codings may simply not be appropriate for handling certain matters that are novel but that nevertheless are allocated those old codings and activate the old rules of processing despite the fact that they arise in qualitatively different contexts as, for example, a problem in quantum field theory, such as was never encountered in the development of the coding derived from gene or cradle and with respect to which that coding is plain wrong. Quantum field theory would then be confronted with, and frustrated by, what in effect would be an unrecognized and undetectable and therefore unspeakable prejudice on the part of the human brain, a prejudice of which, by its nature, it cannot be disabused, and which would not permit the necessary train of thought to be undertaken. Perhaps all that would be necessary to break the logjam would be to recognize the possibility that, just occasionally and under certain special circumstances, blue is red. This was perhaps what Georg Christof Lichtenberg meant when he said:

If an angel were to tell us something of his philosophy, I do believe some of his propositions would sound like $2 \times 2 = 13$.

Although we pride ourselves upon our rationality, and when we can codify that rationality, as in arithmetic, will under no circumstances depart from it, it is salutary to recall the lesson that many psychologists have taught us, namely, that illogical thinking and irrational behavior are more common than not in human beings when they are not operating under one of the standard codifications.

It is also salutary to recall that the great codifications themselves, including even arithmetic, must remain incomplete. Thus the famous and devastating demonstration by Kurt Gödel that any system that is sufficiently complex to contain arithmetic must also contain truths that cannot be proved within the system. Such proof must come from outside the system, in the jargon, from a metasystem, which obviously leads to an unprofitable infinite regression. Although I do not think that he had anything as formal as this in mind, we should also recall Richard Feynman's remark in his Nobel Prize lecture of 1965:

A very great deal more truth can become known than can be proven.

It is the job of philosophers to dig down to such roots of the organization of our human understanding and, as far as they can, to identify its deep assumptions and prejudices. I was once sitting at dinner across the table from the philosopher Gilbert Ryle who was talking to a mathematician who was explaining to him some simple algebraic procedure. The mathematician said:

"Let the number of sheep be x."

Gilbert Ryle said:

"What if it isn't?"

I am saying that the human brain has built-in rigid assumptions and prejudices, inflexible codings, inaccessible to our logical, psychological, and philosophical analysis; that those rigidities forbid the thinking of what they, by their existence, define as unthinkable thoughts just as surely as the taking of other thoughts is impossible because of the brain's inability to establish necessary new codings; and that this must insist that we define and understand the world, not in its fullness, but only in terms of the shadows on the wall of the cave. It cannot be otherwise.

As René Huyghe writes:

Here is perhaps the greatest conquest of the intellect, that it should be able . . . to go beyond itself in order to study itself, know its own frailties, its limitations and its blindnesses.

This tale of human bankruptcy can be told more gently and perhaps more positively, certainly more comfortably. Thus Gilberto Bernardini reports that Wolfgang Pauli said:

. . . the human mind and the objects that we see and that we discover belong to the same perennial order.

Bernardini himself continued:

. . . it was an expression of faith in this congeniality, harmony, that has to be between a man who is trying to investigate Nature and the answer that Nature is supposed to give.

It is the poets rather than the scientists who have, long ago, realized the essentially incomplete grasp that our minds can gain of truths reserved to those not so limited.
From Matthew Arnold:

Mind is a light which the Gods mock us with
To lead those false who trust it.

From Sophocles:

To know is the prerogative of Gods, not men.

But there is occasional comfort of a kind:
Christopher Fry:

. . . Between
Our birth and death we may touch understanding
As a moth brushes a window with its wing.

Of course, we retain our innocence. Slightly to paraphrase George Santayana:

Most questions which it is beyond Man's power to answer do not occur to him at all.

Leaving my talk of madmen and madwomen who can think unthinkable thoughts and returning to our normal everyday world of sane people, that is to say, you and me but not necessarily him or her, unaware of the detailed functioning of our brains and recognizing only input and output: how do we think?

"Top-Down" and "Bottom-Up"

We may distinguish broadly between two extreme modes of thought intended to lead to the solution of the kind of problems with which I am chiefly concerned here. The first we may call the "top-down" and the second the "bottom-up."

Top-down approaches start from a few received attributes such as the natural numbers, a few fundamental laws such as conservation of energy and momentum, a few axioms, and the idea, the obsession perhaps, that the only way to describe the natural world is by measuring things in terms of numbers; they then work out everything from that. But we must realize that the few postulates of the top-down approach and the limited range of possible concepts that such an approach implies restrict the view that we can form of systems that we may attempt to describe or understand through such an approach.

The other of my extreme dichotomy of our approaches to a description of, and an accounting for, the natural world, the bottom-up, is an essentially biological, evolutionary approach in which experiences are stored, compared, and analyzed for common elements from which general empirical conclusions can be drawn to guide future expectations and form the basis of decision making. Those expectations and that decision making then result in actions and further experiences that may modify the system to its advantage or disadvantage. This is evolution.

Numbers. When we consider the bases of scientific thought we begin with Joseph Marie de Maistre around 1800:

The concept of number is the obvious distinction between the beast and Man.

Indeed, extreme conclusions can be drawn from the top-down approach, as when Arthur Eddington maintained in 1938 that from the mere prejudice that the natural world must be described in terms of numbers inevitably flows the conclusion that:

. . . there are 15 747 724 136 275 002 577 605 653 961 181 555 468 044 717 914 527 116 709 366 231 425 076 185 631 031 296 protons in the Universe and the same number of electrons.

I will not trace Eddington's reasoning: it fills a book, but it is worth bearing Eddington in mind to illustrate how intimately our conclusions about the "objective" natural world may be influenced by the prejudices that lie behind our questioning of it.

Feeling and the Rules of the Game. We know there are many things in the world of our human experience that are *not* measured in terms of numbers. For example, we do not express our reaction to a Beethoven quartet by counting something; we appreciate it in terms of its emotional, nervous, and spiritual impact, which is assimilated directly into our experience of the human condition. Counting is wrong in a case like that; there are levels of intensity but they are not quantitatively objective levels. Similarly in our experience of compassion, we are simply moved and although the compassion may be the greater or the less, we do not quantify the degree of movement. For these centrally important things, numbers are just wrong: we feel, we do not count. And yet for the natural world we count but do not feel. Why is that? Is it wholly right? Does not the fact that all our knowledge of the natural world has to come to us ultimately through that same personal experience as expresses itself in essentially nonnumerical terms imply, perhaps to only the tiniest degree, that our understanding of the natural world cannot be complete just in terms of numbers? We count and know, we feel and know; there must be a connection; there can be no completely insulated and noncommunicating compartments within our experience, within the totality of our worldview.

There are certainly those who will argue that love, hate, and compassion, those qualitative, powerful nonnumerical agencies that I am using as a marker for the possibility of a nonnumerical component to the natural world, are themselves merely the complex end product of elementary physical laws and are not factors that should, in some sense, be added to those laws in order to illuminate them and extend our understanding. That counterview was expressed very strongly by Richard Feynman who wrote:

Everything is made of atoms . . . everything that animals do, atoms do . . . there is nothing that living things do that cannot be understood from the point of view that they are made of atoms acting according to the laws of physics.

He was, perhaps, going a little bit beyond what he could actually demonstrate. Feynman is making a statement of faith that draws a boundary to contain the rules of the game. But a boundary to keep our understanding in or to keep our understanding out? As Wittgenstein writes:

If I surround an area with a fence . . . the purpose may be to prevent someone from getting in or out; but it may also be part of a game and the players be supposed, say, to jump over the boundary. So, if I draw a boundary line, that is not yet to say what I am drawing it for.

The quantitative, numerical laws of physics themselves indeed draw the boundary for the present rules of the game of our understanding of the natural world, but we much watch carefully for what may jump over the fence from the other side.

Perhaps Sylvia Plath put it a bit too strongly:

Physics made me sick the whole time I learned it. What I couldn't stand was this shrinking everything into letters and numbers.

Superstition and the Irrational. We must keep a carefully open mind. If we aim, grandiosely perhaps, but in the most broadly questing spirit, at a synthesis of all aspects of our existence and awareness, we must not reject certain things just because we think they are not scientific; that would be to throw out the baby with the bath water. We must not ignore what I might call the reality of the unreal, the reality of the supernatural. Feynman again, speaking in 1964 at celebrations in Pisa marking the fourth centenary of the birth of Galileo Galilei, put it with typical perspicuity and incisiveness when, with his mind on any one of our popular daily newspapers, he imagined Galileo saying today:

"I noticed that Jupiter was a ball with moons and not a god in the sky. Tell me, what happened to the astrologers?"

The astrologers are indeed still with us and doing very well too, with, not so long ago, notorious favor in the highest places. Many times as many people believe in astrology as believe in quantum field theory. Why do we not therefore democratically accept astrology and reject quantum field theory? I do not need to answer my own silly question but I pose it to emphasize that there is evidently something in astrology in that it seizes the minds of so many people and that if we could understand what that something is we might find in it one of those nonquantitative whispers through which the natural world might make itself additionally manifest to us.

It is only with the emergence of modern science in the seventeenth century that distinction has been made between what we would today consider to be the rational and the irrational. Newton more than anyone else brought about that emergence and showed that the physical world might be symbolically represented and that those symbols might then be mathematically manipulated in alternative and different but equivalent ways, by geometry or by calculus for example, and the same conclusion gained, thereby demonstrating the internal integrity of such a worldview. That is well remembered and Newton is rightly honored for it. But what I want to insist on here is that Newton was equally devoted to alchemy with its own set of what we now regard as nonsensical symbolic representations that led to no mathematical manipulation and to no demonstration of internal integrity. Yet, for Newton, alchemy had a parallel relevance to mathematical physics in building up his overall picture of the way the Universe worked. For him the symbols of alchemy were as real and as right as mass and acceleration.

The irrational, the nonnumerical, has, understandably, slipped out of scientific method; it has been excluded by the rules of the scientific game: quantitativity, reproducibility, and so on. I am just entering a cautious and gentle reservation to the effect that, just possibly, we may have been a little too hasty. Because alchemy, for example, proved to be nonsense when today's chemistry came along, that does not prove that we should dismiss, in principle, as irrelevant to the natural world, everything that cannot be converted into digital output. Remember Francis Bacon:

There is a superstition in avoiding superstition.

I find in the response of most scientists to matters such as I am now discussing a close parallel with the attitude of some people, such as the inhabitants of the Chukchi peninsula, the northeast extremity of Asia, who become nauseous at the sight of objects from outside their own culture.

Mathematics. Something related to the use of numbers as the basis for our understanding of the natural world is our use of mathematics. Most thinkers about physical structures, from the classical Greeks through Galileo, who said:

The book of Nature is written in the language of mathematics.

and Newton to the present day, have considered that the Universe is, in the most fundamental sense, mathematical and can be described by equations relating continuous variables such as position and time. Many scientists, including Paul Dirac as we shall see later, almost seem to regard humanity as the servant of mathematics rather than the other way round. But it is an unprovable assumption that mathematics rules and is the language in which all approaches to understanding must be couched.

Even within mathematics itself there is doubt about the basis that we conventionally use in developing our physical ideas, namely, the continuum of numbers: it is a fact that not even a straight line can be described in a way acceptable to mathematical logic. Hermann Weyl wrote:

> . . . belief in this transcendental world of mathematical ideals, of propositions of infinite length, and of a continuum of numbers taxes the strength of our faith hardly less than the doctrines of the early Fathers of the Church or of the scholastic philosophers of the Middle Ages.

And John Wheeler wrote in 1988:

> . . . The physical continuum, and with it all the beautiful machinery of physics, is myth.

Even before we begin to question the logical bases of mathematics itself we must question its status in relation to our own nature. Mathematics is not an inherent property of humanity; even the business of counting has to be acquired rather painfully; some primitive peoples have difficulty in counting beyond 2 and have different words for the same number of different things; ten thousand years ago the Sumerians had two different "2"s for 2 sheep and 2 goats and did not recognize that they could be added together to make 4 animals. As perhaps a faint echo of this do we not ourselves, to this day, speak of a brace of pheasants or a brace of pistols but never of a brace of pencils?

Mathematics and mathematical logic are not part of our deep natural thought processes and codings; they are a synthetic categorization of our experience; our minds do not naturally work that way. Mathematics and logic have rather the status of tools or crutches. We need them because we are not naturally endowed with them. We need to construct logic because our minds are not themselves logical. As Richard Gregory has pointed out, if our fingers were like screwdrivers we would not need screwdrivers. But we have no reason to suppose that the ways in which our minds do work are fully expressible in mathematical terms: there may be a mismatch between mathematics and our full mental capacities; mathematics may be an incomplete way of inputting the natural world to our minds and outputting our understanding of it; logic and mathematics may conceal, as well as reveal, truth. Ambrose Bierce defines it well:

Logic: the art of thinking . . . in strict accordance with the limitations . . . of human understanding.

We must also recognize that mathematics is not itself a unique and coherent structure. The sort of mathematics that we employ to describe the natural world determines and limits the sort of account that we can give of the natural world. The common forms of mathematics have arisen largely from our acquaintance with, and our wish to describe and control, the world of ordinary experience and the extrapolations of it to which our researches have progressively led. Mathematics, to this degree, has had this evolutionary aspect. But just as we continue, by analogy, to use the words we have derived from our everyday lives to talk about Nature on scales of space and time that are inaccessible to our senses, so we continue to use the mathematics that has developed from our everyday experience to describe the workings of Nature in those remote domains.

We do not know that this is best; it may well be that some other sort of mathematics, perhaps already known, perhaps not, will be more suited. Thus, in recent years, topology, the theory of numbers, the theory of knots, and so on, which were developed as abstract intellectual exercises without thought of practical application, have been found more useful for addressing certain physical problems than the mathematics previously used. New domains of physics engender new forms of mathematics, as has happened many times; today particle physics and other branches of science are encouraging new mathematics whose development may lead to a recasting and extension of our basic understanding. Already much more new mathematics has arisen in this way, of the greatest interest in its own intellectual right, than has been applied. The image of mathematics as a whole as a vast space within which exploration has taken place only along filaments or thin sheets of understanding is probably a good one. It is very difficult to take a leap into the totally unknown depths of those unexplored regions, but it may be there that the best tools lie for tackling problems already well defined experimentally both in the fields of the very small and the very large and also in the fields of high complexity. The sort of mathematics we are now employing may simply be inadequate for the problems in question.

Further reservations about mathematics arise from the possibility that ultimate and "correct" mathematical structures for describing natural phenomena may be, in the technical sense, noncomputable and have inaccessible implications and consequences or may involve, again in the technical sense, in-principle insoluble equations. I leave the matter with Bertrand Russell:

Physics is mathematical not because we know so much about the physical world, but because we know so little: it is only its mathematical properties that we can discover.

Science has always worked in these two modes. First, there are the great grand axiomatic top-down procedures that with great conciseness tell us everything—and therefore nothing, as Henri Poincaré remarked. Such procedures are often rigorous in theory but useless in practice because they run into the combinatorial explosion. A good example is the Zermelo-Borel-von Neumann algorithm, which constitutes a complete theoretical solution to the game of chess such that a computer programmed with it would infallibly win every time. This algorithm is very compact and would occupy no more than 10^3 to 10^4 bits of store, but today's fastest computer would take about 10^{90} years to select a move. Such top-down theories are the basis of our understanding but little more; in practice they are useful for simple problems but not for the great majority of those that one encounters in real life: one solves the Dirac equation for the hydrogen atom, and this shows that it is right, but you do not let it loose on the atom of lead because the answer would never come out at the other end.

Second, there are the bottom-up approaches that are pragmatic, syncretic patchworks that build upon experience and on what is found to do the job, conforming to the grand imperatives of conservation theorems and so on, always casting a glance over their shoulders at their top-down masters, and hoping for an approving nod, but in the meantime content to run the errands and do the dishes.

Although I have presented top-down and bottom-up as a dichotomy, they are related in that the bottom-up approach to a problem, when pursued sufficiently far, perhaps through models and analogies, namely, heuristic devices with which the mind, because they are closer to household experience, feels more at home, may lead to a grand rationalization when it is realized that all the bottom-up elements can themselves be derived from, or succinctly expressed through, a single statement that is their top-down synthesis. Thus Brian Pippard writes:

I think that history shows that the imagination needs these props. Few can build without scaffolding; in Maxwell's equations and Einstein's relativity what we see is the final result of a long process, after the scaffolding has been removed. Even Einstein in his quantum theory developments was unashamedly guided by private models of an as-yet-unobserved atomism. . . .

The Patchwork Quilt of Academic Disciplines. To a considerable degree what is regarded as top-down and bottom-up depends on the disciplinary area concerned. Academic study divides itself, by and large, into contiguous areas within any one of which ends are pursued by a restricted range of intellectual technologies. These technologies join smoothly onto the corresponding technologies of the adjacent areas of the patchwork quilt of knowledge such that the bottom-up methods of one become the top-down methods of the area next down the academic hierarchy while its own top-down methods are the bottom-up practices of the academic area next above. It is only rarely that the top-down approach of one area relates to the bottom-up practice of the area next below it and so on. In other words the fundamental concepts of one area are not of practical use in solving the problems of a remote area of academic concern.

Consider just one example of a chain of related academic areas: from the fundamental top-down laws of physics (such as we presently consider them to be although being careful to recognize their provisional nature) we define the basis for our account of atoms and of simple molecules; by this time we are already using the bottom-up methods of configuration mixing and molecular orbitals. At a rather low level of molecular complexity the "fundamentally based" approach of the physicist gives over to the more empirical and pragmatic methods of the chemist, who uses as top-down input the bottom-up methods of the physicist; for the chemist configuration mixing and molecular orbitals are top-down. The chemist arrives at bottom-up methods of bond strengths and ball-and-strut models. The chemist in turn passes the torch to the molecular biologist, thence to the geneticist, to the ethologist, to the psychologist, to the sociologist, to the political scientist, and thence to the psephologist, the student of election results. At each disciplinary interface there is a meeting of methodologies and an authentic and academically respectable sharing of knowledge and of tradition as bottom-up is traded for top-down. But can we therefore say, on the basis of Coulomb's law and the Schrödinger equation, from which the chain starts out at the level of the fundamental physicist, who will be the next President of the United States? No, we cannot, nor must we try. The context of each study defines the nature and meaning of knowledge and understanding in the terms of that study; the intellectual tools we bring to bear to convert knowledge into understanding must be appropriate for the job, not too fine and not too coarse: in those far-off days when wristwatches were mended, one did not do it with a pickax nor does one tear up the pavement with a pair of tweezers.

So for me here: as I explore our relationship with the Universe, I must seriously question whether our tools are appropriate for the job; whether our technologies of thought are not, perhaps in a fundamental sense, inadequate to specify the top-down of the Universe or even whether quantum electrodynamics with its electroweak extension, quantum chromodynamics, and general relativity, which we are hastily attempting to meld together into that top-down "theory of everything," themselves constitute even an adequate bottom-up starting point; and whether there might not be certain missing but essential ingredients without which understanding must remain elusive.

Forces, Laws, and Particles. There is no certainty that the four forces of Nature, the four interactions, of which we are aware, namely, gravity, the strong, the weak, and the electromagnetic, are all that there are. There may well be others, perhaps many others, that act only feebly under the conditions of our everyday Universe but that were of greater, perhaps of dominant, importance under the unimaginably different conditions of the earliest moments of creation that I shall address later. Similar uncertainties must attach to the entities upon which the forces operate: we are familiar with many forms of the organization of matter that we divide into families of particles; some families are acted upon by all four forces, others are choosy and ignore some of the forces completely; the only known force that acts on all particles is gravity. But there may be other forms of the organization of matter, other families of particles, that we do not know about because they interact only extremely feebly with the particles of our normal experience either in the kitchen or in the laboratory. But those other families may interact strongly with each other through forces of which we are not aware because they operate only extremely feebly within and between our familiar families, as well as between those families and the new ones of my hypothesis.

Thirty years ago I permitted myself to write:

Perhaps there do indeed exist universes interpenetrating with ours; perhaps of a high complexity; perhaps containing their own form of awareness; constructed out of other particles and other interactions than those that we know, but awaiting discovery through some common but elusive interaction that we have yet to spot. It is not the physicist's job to make this sort of speculation, but today, when we are so much less sure of the natural world than we were two decades ago, he can at least license it.

And in today's today, after a further three decades of increasing unsureness, I begin to feel that the physicist has a correspondingly increasing responsibility for drawing attention to this possibility; the mind must be kept attentively open.

The forces operate between the particles and the laws of Nature tell you how they do it. But just as there may be many families of particles and interactions of which we are not yet aware, but without knowledge of which there can be no theory of everything, so there may be many laws that remain undiscovered. The very meaning and status of the laws of Nature is unclear. Some laws, such as the conservation of energy and momentum, seem to be very good and of universal applicability, but others apply only to certain of the interactions, such as the conservation of parity, the mirror symmetry of Nature, which works very well, perhaps perfectly, for the electromagnetic and the strong interactions but which appears to be, in our household experience, totally violated in the weak interaction. Are all laws, in fact, only approximate, being only a question of degree? I think it was Ralph Waldo Emerson who said:

Everything that God has made has got a crack in it.

We do not know.

No pLaws At All? Are there, in fact, any laws of Nature at all? Presumably one cannot have a law of Nature in the abstract: there must be a set of entities whose mutual disporting is codified through what we call the laws; we cannot contemplate the existence of laws of Nature without a Nature within which that existence manifests itself. But as we shall see later, when the Universe came into existence there was probably a high degree of arbitrariness in the way in which it evolved into the form in which we ourselves now find it; that is to say, a high degree of arbitrariness in the way in which the laws themselves emerged, together with their nuts and bolts of the natural constants.

It may be only by chance that there appear to be any laws at all, that the most probable situation would be a wholly irrational Universe, without cause and effect, within which order, and so life, could not have arisen. The laws of Nature, as we know them, may therefore not really be laws in the sense in which we usually use that term but just our codification of an accidentally rational behavior on the part of the Universe. There appear to be laws of Nature because it is only in an orderly Universe that life and intelligence could have emerged. But I begin to trespass upon a later chapter of my argument and so I shall leave it there for the moment.

On a more technical level, some thought has recently been given, particularly by H. B. Nielsen, to the possibility that there are in fact no fundamental symmetries at the very high energies characteristic of the earliest instants of the Universe, or that, what is equivalent, there is a chaotic and random superposition of all possible symmetries and therefore of all possible laws. And then, as the Universe ages and cools, in the manner that I shall discuss in some detail later, some of the tangled mass of symmetries become irrelevant until, by a kind of process of natural selection, those symmetries best suited to the governance of a large-scale cool Universe, such as we inhabit, remain and we grandly call them "the" laws of Nature as though they, and only they, had been there all the time.

So while searches for a self-consistent theory of everything are fine and prodigious intellectual exercises, it may be that they are misconceived in the sense that they aim at a top-down based on an incomplete bottom-up. It may be better to hearken to Hermann Bondi:

> I regard it as an essential of any scientific theory to have room for putting in what one does not know yet.

Memory. An essential element in the operation of our thought processes is memory, which is usually, but not always, regarded as part of mind. The idea that we have two distinguished aspects, body and mind, has been, and continues to be, something of a straitjacket for all except the extreme behaviorists and their epigones in the field of artificial intelligence. The making of our judgments depend on our bringing together, out of our memories, the appropriate constellation of information, there stored, with the appropriate degree of simultaneity and, very probably, although we do not know for certain, in the appropriate sequence. It is this bringing together that occasionally leads to what I might term the bottom-up "Eureka" or "Aha!" that we experience when a specific problem is suddenly solved and that is to be contrasted with the top-down "Aha!" that comes when we realize how a single simple idea illuminates a vast range of phenomena and processes; to this I shall return.

There have always been attempts to describe memory through some sort of physical analogy. Plato reports Socrates as using the metaphor of a wax tablet that

> . . . we hold to our perceptions or thoughts so that it receives the impression of them as from the seal of a ring so that we know what is imprinted as long as the impression lasts.

The other major hypothesis was expressed by Aristotle shortly afterward, namely, that memory is not a static record such as a wax tablet or a computer tape but is rather a continuously maintained relationship between the pneuma, our vital spirit, and our flowing blood, more like a computer's active memory—although neither Socrates nor Aristotle actually used those analogies.

In the English language we follow Aristotle and say "I know it by heart." Other languages, for example, Dutch, rather follow Socrates and say "I know it out of my head." In English, of course, the head and heart have distinct roles, as revealed in C. D. B. Ellis' little verse:

> I laid my hand upon my heart
> And swore that we would never part;
> I wonder what I would have said
> Had I laid it on my head.

But that is another story.

John Zachary Young writes:

We have not yet learnt enough about ourselves, and our brains and our language, to overcome entirely the difficulties presented by the dual concept of body and mind. . . . Nevertheless we can begin to understand that the "immaterial principle" that operates within us is the manifestation of the information coded within our physical brains. Information is not material . . . (but) it can be carried and manifested only by material systems, whether active or passive. The pattern that is maintained by living things is certainly not a static structure but a continually active system interchanging with the environment. We can usefully think of the "immaterial principle" that seems to pervade life as the encoded information that directs those ordered activities so that life continues.

The view, now almost universally held, that the brain stores its information in coded form and performs symbolic operations upon it has been reached slowly and painfully by neurophysiologists over many years and wholly without reference to computer architecture and functioning; yet how easily the structures of the mental and computer systems can be put into relationship with one another, at least at the most fundamental, if not yet the operational, level. It is this, recalling vividly the little girl's second effort at drawing a square, the one that looked more like a beetle, that emboldens some to say that computers may become a significant supplement to our own minds.

Selectivity and Censorship. A tremendously important factor for our present consideration is the selectivity of human memory. When we remember we do not bring back everything our brain contains but rather a digest, from the whole, of those things that we unconsciously consider to be relevant to the project in hand. As Frederic Bartlett said:

Memory reconstructs the past rather than reproducing it.

This selectivity of human memory, the bringing together of those elements of the total store that our internal censor considers to be relevant for the illumination of the problem in hand, implies purpose in the search and recall. The coding of memory must therefore be a tagged coding that also labels purpose or ranges of purpose. This is an important factor in our quest for an understanding of the Universe. Our recall from memory, our examination of the evidence, is subject to this internal censorship and judgment about relevance. This is familiar enough in normal contexts where a problem that has been widely recognized for a long time is suddenly solved, in what immediately becomes a most obvious way, because a previously neglected aspect has pierced the veil of censorship. As Seneca said nearly 2000 years ago:

Posterity will be astonished that such obvious truths had escaped us.

Or as Schopenhauer put it:

It is not so much a question of seeing something new as of thinking what nobody has thought before about something that everybody sees.

Or as George Bernard Shaw:

No question is so difficult to answer as that to which the answer is obvious.

Most probably many people, in fact, have those novel thoughts of which Schopenhauer speaks, have them escape the first censorship, only to be similarly blocked on their way to consciousness as in the phenomenon of prosopagnosia to which I earlier referred. Perhaps Goethe knew something when he wrote:

Everything has been thought of already, the only problem is thinking of it again.

And, he should have added, "knowing that you have thought of it." Even Coco Chanel might have known something when she said:

Only those with no memory insist on their own originality.

Such selectivity and censorship stand in the way of our solution of everyday problems in closed and familiar contexts. How much more serious are these restrictions when we face the unknown context of the Universe as a whole whose dominating influences may speak to us only through scarcely audible whispers, attenuated by space and time, and of a nature that our brains may not know how to encode.

Before we worry too much about the fact that our mental equipment may not be appropriate for bringing us into comprehension of the unknown Universe we must recognize that there are those, led by B. F. Skinner, who hold that our psyche is just one big mistake, unsuitable for coping even with the world of Galileo, Newton and, indeed, that of the kitchen sink. As Jerome Bruner has written:

. . . in recent years, the most conspicuous voice of psychology has been . . . motivated by the assertion that scientific psychology shows that the human enterprise is altogether wrongly conceived . . . (that we are) muddled by notions like choice, freedom, dignity, intention, expectations, goals, and the like . . . and that human affairs so conceived can be shown to be "wrong" in much the same way as Copernicus showed that the heliocentric Universe was "wrong."

This does not seem to me to be a very encouraging starting point for our attempts to understand the nature and meaning of creation. But whatever our response to the idea that humanity is a bit of a mistake, it is all that we have available. Inevitably, as I have been painfully insisting, we are limited in the responses we can make to the signals that Nature sends us; certain signals, perhaps essential to the task of comprehension, escape because we cannot take them in or because we cannot output their consequences to our consciousness. We are like the epistemological fisherman whom Eddington saw compiling a catalog of all the fishes of the sea but using for his task a net with a rather coarse mesh; when Eddington told him that his net would catch only the big, but not the little, fishes the fisherman replied: "What my net won't catch isn't fish." We must try not to make that mistake—but how do we know that we are not making it?

The walls of the caves of science are thick with Plato's shadows.

Generalizing my earlier remark about the obsessively dominant role that numbers play in our description of the physical world and tacking onto it my remark about our de facto psyche and the possibility that it is, in some sense wrong, or perhaps incomplete, we are only confirming in specific terms that there are indeed prejudices about the natural world truly inherent in us, to the degree that they should not be called prejudices because they constitute part of our definition and we cannot be disabused of them and that these prejudices render us inherently incapable of a full understanding of the natural world because they lead us to insist on attempting to describe it in essentially incomplete terms.

We cannot escape from the unchanging context of our own existence and so cannot objectively examine that context and appraise its role in conditioning our perceptions. As Einstein said:

What does a fish know about the water in which he swims all his life?

Or, as Arthur Miller put the same thought:

The water is in the fish and the fish is in the water. There's no separating the two.

A computer would not suffer from these inhibitions and we cannot build them into it because we are, by definition, unaware of them. So there is a possibility that a computer might recognize regularities or relationships in the course of its symbolic manipulations that are real but that we should reject as gibberish. Perhaps we shall very slowly learn new modes of thought and feeling; perhaps the computer may indeed have a hand in the evolution of our psyche. But it will be slow. We start with John Dewey:

Old ideas give way slowly; for they are more than abstract logical forms and categories. They are habits, predispositions, deeply ingrained attitudes of aversion and preference.

I am giving to each of Dewey's words a more absolute significance and welding it firmly to our intrinsic definition.

The Advance of Our Understanding. From the basis of this recognition of our prejudiced inadequacy I now turn to a central point of my thesis in this book: the mechanism of our discovery of new principles and concepts. How does science, more generally our account of the Universe in which we live, advance?

Advance takes the form, in part, of an extending and refining of our experimental data and of the concepts and theories that have stood us in good stead, thus far, for the synthesis and rationalization of those data. But, in major part, advance consists in the introduction of new concepts and new paradigms; the revolutions of Thomas Kuhn. In this there is a proper spectrum of view, from the activist revolutionary who immediately seizes upon any new idea and throws out of the window everything that preceded it without regard for antecedent triumphs of understanding, to the conservative who is so reluctant to accept a new concept that he or she resists to the point that the consensus of peers regards as sheer obscurantism. It is good that this spectrum exists; scientific advance is best when its progress is one of measured tension and is at its least enlightening, and it is most likely to lead to disaster when it resembles a rush of Gadarene swine or a surge of lemmings. I do not give examples; they spring all too easily to mind. Advance is best achieved by this constructive tension within the scientific community; the scientific community is always willing, indeed eager, to accept new concepts but only when their clear superiority to the old has been demonstrated, when, in short, they are obviously called for.

Advance should not be equated with fashion. As E. Michel Cioran says:

A philosophical fashion catches on like a gastronomical fashion: one can no more refute an idea than a sauce.

Atoms. The history of the atomic hypothesis (as some, of classical bent, indeed still call it) is a good case history. From the days of Leucippus, Democritus, Epicurus, and Lucretius through to the early nineteenth century, atoms were philosophical abstractions about whose reality there was nothing but anecdotal evidence; you paid your money and you took your choice. Then came John Dalton who, in 1808, showed that the evidence of chemistry could be rationalized through the existence of atoms, which then, for some, began to take on a cautious measure of inferential existence. Through the nineteenth century more and more phenomena were investigated and more and more developments of the atomic theory took place, particularly within the kinetic theory of gases, all consistent with the atomic hypothesis; a substantial body of belief built up. But, of course, nobody had ever actually seen an atom and nothing but essentially inferential and statistical evidence for their existence was available. But there was also plenty of resistance to the idea and the hypothesis remained a hypothesis.

Then, in 1906, Einstein, by his analysis of Brownian motion, showed that the incessant jiggling of pollen grains suspended in water, known since 1827, was quantitatively just what one would expect if it were indeed due to the random bombardment of the grains by molecules of water precisely according to the dictates of the atomic hypothesis. Almost all the conservatives were then convinced, including Wilhelm Ostwald, whose reservations on the ground that atoms were not necessary for his understanding of chemistry were overcome. But atoms had not been seen as such, and their existence, even though the phenomenon of Brownian motion enabled them to be counted, remained for a few but a hypothesis, no matter how successful. Thus Ernst Mach, the physicist-philosopher, held out against the reality of atoms. In 1883 he had written:

Atoms cannot be perceived by the senses . . . they are things of thought . . . a mathematical model for facilitating the mental reproduction of the facts.

But Mach's stance was at least partly on the ground that we were being perhaps excessively simplistic in describing the invisible atoms in household terms whose appropriateness was established only for macroscopic objects:

Thereby we suppose that things which can never be seen or touched and only exist in our imagination and understanding can have the properties and relations only of things which can be touched. We impose on the creations of thought the limitations of the visible and tangible.

Mach held out to the end against attributing to atoms a literal reality of which, as he thought, we could never be assured. He died in 1916 but late in 1913 conceded in conversation with Einstein that atoms were a useful physical hypothesis, but he did not grant them real existence. Nowadays we not only see individual atoms in the electron microscope, we see their individual internal structure through the tunneling electron microscope and infer it as clearly by holographic methods as we do that of ordinary macroscopic objects. It would be nice to know what Mach thought about that.

New Concepts. I have labored this illustration somewhat because it is itself a paradigm for the way in which new concepts arise and are processed and accepted. The lesson is that we should adopt and move to a new concept, a new layer of understanding, only when we are forced to do so and not simply because its exploration may be intellectually challenging and exciting. We otherwise face the dilemma posed by Roger Fry:

> . . . whether a theory that disregarded facts would have equal value for science with one which agreed with facts. I suppose we would say No; and yet so far as I can see there would be no purely aesthetic reason why it should not.

Thus we were forced to belief in the atom long before we had actually seen it. Then, within the atom, we were forced to belief in the atomic nucleus the seeing of which and of its constituent neutrons and protons requires some, but not great, extrapolation of methods that we accept as enabling us to see objects that we can also touch, in particular the analysis of diffraction patterns. Then, within the proton, we were forced to the quarks by different methods of kinematic inference that do not relate to seeing in the classical sense at all because part of that inference is that quarks, in fact, have no structure to be seen. And yet we believe in them implicitly (in both senses of the meaning of that word) despite the fact that they have never emerged as individuals from their nucleonic residence.

Analogy. In all this we are assuming, of course, that Nature continues to work by direct analogy with the objects of our everyday experience and that we are not, by making that assumption, just constructing a fantasy world although indeed we may be. Both David Hume and John Stuart Mill held that all reasoning depends on resemblance or analogy; as Samuel Butler put it:

Though analogy is often misleading it is the least misleading thing we have.

Our appeal has to be to the plausibility of the story we are telling and to its internal consistency: if you know a better hole, go to it. This faith in the continuing applicability of analogy, and in Man's representation, or construction, of Nature by those means, which I shall touch upon again later, was expressed by Herman Melville through the mouth of Captain Ahab:

O Nature and O soul of Man! how far beyond all utterance are your linked analogies! not the smallest atom stirs or lives in matter, but has its cunning duplicate in mind.

Elaboration. We are sometimes forced to new concepts by an aesthetic revulsion against a rococo elaboration that we can no longer credit. Thus when the elementary particles had become, by 1960, such a vast slough that revulsion against the idea of their elementariness was almost universal (not quite universal, as was proper; conservatives are always useful as I have remarked), the community immediately welcomed their collective description through the Gell-Mann—Ne'eman SU(3) symmetry scheme, which saw them as alternative aspects of each other. This symmetry scheme later led to the concept of the quarks. Today there are those who feel that we have almost as revolting a proliferation of quarks and other basic particles as we had in 1960; they are therefore promoting the idea that the quarks themselves are not elementary but have a substructure of yet more basic entities; but the revulsion is not yet great enough to push the community insistently in this direction once more. Note, however, that I am here asserting the very human quality of revulsion as one of the criteria through which we build up our view of the Universe.

Unification. Another powerful motivation for moving to new concepts is a wish to achieve unification, the simplest single starting point, in our description of Nature. I will have a lot to say about this later but here just remark that this leads us into realms where direct verification is out of the question and where inference and circumstantial evidence are all that we can look to. We do not like this but we should not be surprised that it is the only way. But it does mean that we shall then insist on describing Nature through the only means available to us, Captain Ahab's "linked analogies," and will be naked before Mach's stricture.

Theory and Experiment. Returning to our more general examination of how science advances, it is useful to start with a remark from one who, more than anyone else this century, has been responsible for that advance, namely, Einstein:

> The scientific theorist is not to be envied. For Nature, or more precisely experiment, is an inexorable and not very friendly judge of his work. It never says "Yes" to a theory. In the most favourable cases it says "Maybe," and in the great majority of cases simply "No." If an experiment agrees with a theory it means for the latter "Maybe" and if it does not agree it means "No." Probably every theory will someday experience its "No"— most theories soon after conception.

Another remark of Einstein's is also useful here as reinforcing the point I made earlier about our need to be forced from paradigm to paradigm rather than flit by fashion's diktat. Speaking of his general theory of relativity:

> I am anxious to draw attention to the fact that this theory is not speculative in origin; it owes its invention entirely to the desire to make physical theory fit observed fact as well as possible. The abandonment of a certain concept must not be regarded as arbitrary but only as conditioned by observed facts . . . the justification for a physical concept lies exclusively in its clear and unambiguous relation to the facts that can be experienced.

A striking feature of scientific advance is the psychological nature of our own involvement in it: first we long for our theory to be right and then, when it seems that it might be, we long for it to be wrong because its being wrong would show that we are not yet at the end of the road. As Sheldon Glashow has said:

One strives to find data to destroy theories that we have struggled to make.

The end of the road is always to be feared; what was once wonderful becomes sadly commonplace in science just as William Butler Yeats saw so clearly in a broader context:

Things thought too long can be no longer thought
For beauty dies of beauty, worth of worth.

So how, in general terms, is science done?

Hypotheses. Science works uniquely through hypotheses. Some are durable and become enshrined in axiomatic structures of alleged understanding, but although they may be graven on tablets of stone (and some indeed have been), hypotheses they remain in all their vulnerability. Roger Coates, in the preface to the second edition of Newton's *Principia,* wrote:

Those who assume hypotheses as first principles of their speculations, although they afterwards proceed with the greatest accuracy from those principles, may indeed form an ingenious romance, but a romance it will still be.

Let us not forget.

You can, with certainty, prove an isolated hypothesis to be wrong, and Karl Popper has taught us, depressingly, that this is the essence of science, but you can never prove a hypothesis to be uniquely right. But a disproved hypothesis can always be salvaged by an *ad hoc* supplementary hypothesis. This sequence can never be stopped because we can never prove the nonexistence of something. Remember the man who said that he would continue to believe in the existence of Water Babies until someone had shown them to him actually not existing. This is rather like the indignant letter that appeared in a Church of England West Country parish magazine, some sixty years ago, at a time when there was fear of Popish plots, accusing the newly appointed Anglican curate of being an agent of the Vatican on the grounds that he had been seen practicing celibacy in the open streets.

So what, in practice, do we scientists do? We work on the basis of some assumedly solid corpus of theory that accommodates the facts brought to light until last week. This week's facts are the new tests to which the current theory is subjected and to accommodate which it may have to be refined. If we can refine the theory without changing its basic structure we are pleased and this enhances our conviction of the theory's rightness. If, however, a supplementary hypothesis has to be injected, we feel uneasy but press on. But at some stage we may be faced by some majestically novel fact that baffles us and that offers no hint of how we might best accommodate it. Our struggles throw up a number of alternative drastic *ad hoc* hypotheses. All are, by definition, right in that they accommodate the new facts because that is what they were invented for.

Judgment and Feeling. How do we choose when we have used up all our scientific criteria, when there are no more facts available or accessible, when we have exhausted scientific method? We are left face to face with ourselves. The only remaining criterion is what seems right to us, not in the head, because the head has done its job of framing the question and its alternative possibilities of resolution, but rather in the deepest seat of human feeling that many from St. John the Evangelist to Oliver Cromwell have held to be the bowels. We suddenly feel in our bowels, with unshakable, unreasoning conviction, that one solution is right. This is the "Aha!"; this is, to continue in ecclesiastical vein, the most intense experience of which humankind is capable, the descent of the divine afflatus. It is a response of our total being: not for me the dictum of St. John Chrysostom who regarded the body as

But the repository of phlegm and spittle.

Nor, for that matter, that of Algernon Charles Swinburne who said

The soul squats down in the flesh like a tinker drunk in a ditch.

Not for nothing do we speak of a gut feeling.

This resolution of judgment is very familiar. It is precisely how we make the choice if, for whatever reason, the choice must be made, between two paintings, two poems, two motets, two courses of action that stir the bowels of our compassion. In all of these situations we sift and exhaust the evidence, we review the totality of our experience, and then "Aha!"; we choose because we *feel*. And so in precisely the same way in science we sift and exhaust the evidence and then choose because we feel. Our psyche, our essential and unquantifiable humanity, must become, at this stage, integral with science itself. Immanuel Kant wrote in the *Critique of Pure Reason:*

Two things fill my mind with ever-increasing wonder and awe. . . . : the starry heavens above me and the moral law within me.

I am simply saying that these are not two things but one. But if our psyche is flawed, and not only flawed but based on incomplete sensibilities, what hope for our ultimate conspectus of the Universe?

Beauty. When we have this experience of the divine afflatus we indeed feel that it impacts upon, or arises from, some intrinsic intensely personal reference point representing our very nature and incapable of further analysis or discussion and that we have discovered it within ourselves. This emotional experience of rightness and its deep and unique relationship with the human spirit was expressed long ago by Plato in the *Phaedrus:*

The soul is awe-stricken and shudders at the sight of the beautiful, for it feels that something is evoked in it that was not imparted to it from without by the senses, but had always been laid down there in the deeply unconscious region.

We attempt to rationalize our emotional feeling and to explain the choice it represents by saying that it is the most beautiful or the most economical and appeal to Keats or to William of Occam respectively. But this is really the wrong way round: we should rather define beauty as what engenders the feeling and leave it at that.

Three hundred years ago Spinoza wrote:

> I ascribe to Nature neither beauty, deformity, order nor confusion. It is only from the viewpoint of our imagination that we say that things are beautiful or unsightly, orderly or chaotic.

If we make beauty the touchstone that guides us toward our understanding of Nature, then we must recognize we are saying that Nature, as well as beauty, is in the eye of the beholder. That, in a nutshell, is the nub of my thesis.

Much has been written about beauty as the most important determinant in science. Beauty is, indeed, often ranked above truth if what is meant by truth is the fitting of the facts: as in Paul Dirac's famous aphorism:

> . . . it is more important to have beauty in one's equations than to have them fit experiment.

Or with Hermann Weyl:

> My work always tried to unite the true with the beautiful; but when I had to choose one or the other I usually chose the beautiful.

Or with Werner Heisenberg recording a conversation with Einstein:

> If Nature leads us to mathematical forms of great simplicity and beauty that no one has previously encountered, we cannot help thinking that they are "true," that they reveal a genuine feature of Nature. . . . You must have felt this too: an almost frightening simplicity and wholeness of the

relationships which Nature suddenly spreads out before us and for which none of us was in the least prepared.

We must be careful how we interpret this apparent emphasis that beauty should be ranked above truth. It is not that at all. What is being meant is that something of great beauty must, you feel, surely be right, but not necessarily right as the solution to the problem that gave rise to the feeling. You feel that it must fit in somewhere and that it would be a great shame if Nature had missed the opportunity to avail itself of such elegance and attractiveness in one or other of its structures; the trick is then to spot which structure. This has happened many times in the history of science. Erwin Schrödinger wrote down a relativistic equation that he hoped would describe the electron, but it did not work because he did not know about the spin of the electron, and it was left to Dirac to fit that in; but Schrödinger's beautiful equation was not wrong: it was correct for a particle such as the π-meson that does not have a spin, and the equation now goes under the name Klein-Gordon. Weyl himself derived what today we would call a gauge theory of gravitation; it was of great beauty, but it was wrong for gravity; it later became one of the cornerstones of quantum electrodynamics.

So beauty that clutches at you in an insistent way must always be welcomed with respectful suspicion. As Alfred North Whitehead said:

Seek simplicity and distrust it.

It was also Whitehead who stated the present matter in an alternative form:

It is more important that a proposition be interesting than that it be true.

We may wonder whether there is within us a kind of hidden hierarchy of considerations that operate in leading us to the "Aha!" that triggers the descent of the afflatus; are some influences and appeals more potent than others? There are those, such as F. S. C.

Northrop, who hold that we possess a kind of instinctual aesthetic sense that tells us what is right and that we therefore lead ourselves toward world pictures that conform most closely to that sense. To caricature this: Nature indeed copies Art. But what Art? That we surely cannot know. In Thomas Mann's *Doktor Faustus* music triumphs over literature: Schopenhauer held that other art forms only aspire to the condition of music; Pythagoras saw the unity of astronomy, mathematics, and music. So perhaps music rules. But then we have Charles Ives, who should know:

There is more to Nature than frequency vibration ratios.

I do not think that the divine afflatus will submit to analysis.

We may value simplicity and beauty and hope they will lead us to the truth, but not all truth is simple and beautiful, or, at least, not all that may be an exceedingly successful surrogate for the truth, as in the basic equation of the electroweak unification that I shall speak of later and that is one of the high points of the theoretical physics of the century; this equation is anything but elegant and succinct, and I believe that not even its parents, Steven Weinberg and Abdus Salam, think that it is beautiful. But when we know more it may grow up into a swan.

Feeling as Touchstone. However you define it, whichever way round you look at it, it must be granted that our ultimate judgments in science are based on personal feeling, not on so-called scientific facts. Of course, these judgments are ultimate only in the sense that they come at the end of a process that has exhausted all other mechanisms. They must themselves be reviewed when new facts come along to enable the scientific method to be engaged once more. It goes without saying that when that happens those afflatus-born judgments, arrived at with total conviction, are often found to be wrong: the afflatus may be divine but in our recognition of it we are only human. We must always remember Francis Bacon's warning:

. . . whatever the mind seizes upon with particular satisfaction is to be held in suspicion.

It is through the afflatus and its well-known little brother, the tingle factor, that we escape from B. F. Skinner's characterization of humanity as a preprogrammed automaton, rigidly molded by our environment and incapable of actions other than conditioned reflexes.

Intuition. On the way to the divine afflatus there are other points at which scientific method uses techniques of dubious scientific or logical validity. The most obvious, and that closest to the afflatus, is intuition. So often, a scientist, when pressed, will admit that he or she did not nibble the way from A to B by a chain of testable reasoning of total clarity but rather knew that B was the right answer. As Henri Poincaré says:

Logic . . . remains barren unless it is fertilized by intuition.

and also:

It is by logic that we prove, but by intuition that we invent.

Sometimes the nibbling route could have been taken but sometimes not; sometimes a leap was necessary to cross a chasm for which no logically constructed bridge could have been provided. There are interesting differences of national style in evidence here: the logical Latins, the mystical Slavs. But it is sure that we must not spend our entire scientific lives nibbling, as Alfred North Whitehead saw:

Insistence on clarity at all costs is based on sheer superstition as to the mode in which human intelligence functions.

The way from intuition to its symbolic codification in the mathematical structure necessary for the expression of that intuition in a form permitting its testing and predictive development is also something of a matter of style. Thus French science, from its standpoint of logic and deductive rationality, looked down its nose at James Clerk Maxwell's early struggles toward a field theory of electromagnetism through which to express Faraday's intuition. Pierre Duhem wrote tartly of Maxwell's

. . . lack of concern for all logic and even for any mathematical exactitude.

and Henri Poincaré, in similar vein, said that when a French reader opened Maxwell's book he experienced

. . . a feeling of discomfort and often even of distrust.

Induction. There is the highly vexed question of induction. Scientists are commonly thought to work by deduction, namely, the inference of something from given premises that must be true if the premises are true. And this is fine for pursuing the consequences of a top-down axiomatic starting point but, in practice, is not of much use for discovery, and discovery is what science is chiefly about. More often, discovery involves induction, which is an inference based on the generalization of experience and which is then assumed to hold universally but, in fact, may not; the premises do not, in induction, entail the conclusion. Thus a chicken that has been fed every day by the same person at the same time may conclude, by induction, that it will be ever thus. But, as Bertrand Russell put it:

The man who has fed the chicken every day throughout its life at last wrings its neck instead, showing that more refined views as to the uniformity of Nature would have been useful to the chicken.

The intuition of the scientist is largely built up inductively but largely unconsciously and, by definition, without its being articulated. But it is through the unexpected contradiction that Nature throws up to the intuitive and inductive expectation that discoveries are made. The only difference between the scientist and the chicken is that the chicken gets it in the neck and the scientist gets the Nobel Prize. The successful scientist carries around a great store of inductive anticipation through which to recognize the unexpected when it occurs. As Louis Pasteur famously put it:

In the fields of observation chance favors only the prepared mind.

Philosophers rage about induction and many of them insist that it is a pernicious and useless concept. I think that this is because philosophers, although they strenuously deny it, long for certainty whereas scientists are happy with a good hint. Rudolf Carnap, in my view somewhat circularly, spoke of humanity's having

. . . an intuitive sense of inductive validity.

And there we should leave it, recognizing that induction, whether one knows it or not, is the way the scientist goes about one's business. Oliver Wendell Holmes put the scientist's spirit very well:

You must see the infinite, i.e., the universal, in your particular, or it is only gossip.

One does not always consciously see the universal in the particular and often the discovery of the universal, through the descent of the afflatus or otherwise, is not seen as related to the antecedent accumulation of facts and the fruitless attempts to marshal coherently and consolidate those facts into a succinct understanding. Einstein wrote that he abandoned hope

of the possibility of discovering the true laws by means of constructive efforts based on known facts. The longer and the more despairingly I tried, the more I came to the conclusion that only the discovery of a universal formal principle could lead us to assured results.

The universal principle came in the form of the special theory of relativity but it came only because of knowledge of, and struggle with, those "known facts." Einstein was also very clear about this when he wrote:

Pure logical thinking cannot yield us any knowledge of the empirical world. All knowledge of reality starts from experience and ends in it.

Jean Paul Sartre was wrong in saying:

Facts and essences are incommensurables; one who begins his enquiry with facts will never arrive at essence.

Fact and Theory. Before we leave facts as the necessary input to theory we must recognize that, in varying measure, facts are themselves theory dependent: they are the product of looking at the world from a particular viewpoint and in a particular context and both the viewpoint and the context represent theory-sensitive selectivities. There are very few totally unbiased, theory-free facts. Things may not count as facts even though they may be, literally, blazingly obvious. Thus the supernova seen by Chinese astronomers in July 1054 went unrecorded in Europe because such things were not there part of the *Zeitgeist:* the supernova did not count; it was not a fact.

It was only our gradual liberation from the medieval worldview that led to the incorporation of what we now call science, with its comparatively open-minded stance, into the common cultural background and to the giving of a finer mesh to Eddington's epistemological fisherman. But with my continuing insistence that humanity must be, as Protagoras put it in the time of Plato, "the measure of all things," I do stress "comparatively open-minded." It cannot be overemphasized that science is not given to us from outside ourselves but is no more than the description and ordering of our experience and that that experience should not, to the uttermost of our ability to recognize it, be subject to a priori censorship.

The Divine Afflatus. The move from fact to essence is always a mixture of induction and intuition. So if I should wish to phrase the experience of the divine afflatus in slightly more technical terms I should say that it is the sudden and unexpected consistent convergence of a number of strands of intuitive/inductive perception.

And there we should leave that too. Except, perhaps, to emphasize that the afflatus arises from the sudden fusion of feeling and reason: convergence of the reason gives rise to the feeling, and this both qualifies and denies Blaise Pascal's dictum, at least in the context of scientific discovery:

Those who are accustomed to judge by feeling do not understand the process of reasoning, because they want to comprehend at a glance and are not used to seeking for first principles. Those, on the other hand, who are accustomed to reason from first principles do not understand matters of feeling at all, because they look for first principles and are unable to comprehend at a glance.

Communication. Something that is complementary to the afflatus is knowing when we have got it right. Not in the sense of the inspirational moment—that is the job of the afflatus—but rather in the workaday sense of having put things into the optimum shape for communication by whatever medium is appropriate. This also is a matter as much for feeling as for anything else. Communication is not just the transmission of facts; it is more the transmission of understanding, which can come about only by engaging the recipient in the process. Perhaps artists understand this better than scientists: any good painting invites audience participation, some even to the point of leaving a beckoning place, literally or in terms of a contribution that the viewer feels that he or she must make to the meaning of the work or to extract the meaning from it. In music too: there are some sorts of chamber music to which I always find myself supplying an unwritten bassoon part in my head; it is not that the bassoon part should be there but that by putting it in I involve myself and so get more out. Some artists have felt that this beckoning vacancy should be carried to the point that Picasso indicated when he asked

"Of what use is a finished picture?"

(He was speaking of his *Acrobat and Young Harlequin,* which in 1988 sold for some $35,000,000, so his point answered itself in a very practical way.) Others have felt that there was some criterion to which appeal could be made. Thus Renoir said that a picture was finished when he wanted to slap its bottom, and Cezanne said that it was finished when it "clasped hands." (Much of Henry Moore's early work, both graphic and plastic, has the figure literally clasping hands.)

Artists have always recognized, in differing measures in differing times, that their work must be matched to the capacity of those who, it is hoped, will appreciate it. Similarly, in science, we respond only to those aspects and phenomena in Nature's grand exhibition gallery that we are fitted to appreciate and grasp, and reject or ignore those whose significance escapes us. We are our own filter, our own epistemological fisherman.

In science as in art it is also important that the work engage other consciousnesses, and the judgment about how best to do this is also a form of creativity: what, if only by implication, is best to leave as unfinished business for the concerned reader. Perfection, in science as in everything else, is sterile.

Starting Off. How do we start it all off? We should not examine and question the mechanisms by which science advances without examining and questioning the propositions from which it starts. I am not here speaking of those deeply rooted and inaccessible factors that constitute our definition and inarticulable prejudices but rather those things that could be said about the foundations of scientific or philosophical inquiry but that are more usually left unsaid. As Francis Cornford put it:

If we look beneath the surface of philosophical discussion, we find that its course is largely governed by assumptions that are seldom, or never, mentioned. I mean that groundwork of current conceptions shared by all men of a given culture and never mentioned because it is taken for granted as obvious.

Alfred North Whitehead put it similarly:

In every age the common interpretation of the world of things is controlled by some scheme of unchallenged and unsuspected presupposition: and the mind of any individual, however little he may think himself to be in sympathy with his contemporaries, is not an insulated compartment, but more like a pool in one continuous medium—the circumambient atmosphere of his place and time.

Where we start is often left undefined, but if error is there, then the whole edifice is built upon shaky ground, so there should be some mechanism of authentication. But this is very difficult. There has to be some unquestioned beginning. As Ludwig Wittgenstein puts it:

If you tried to doubt everything you would not get as far as doubting anything. The game of doubting itself presupposes certainty.

Wittgenstein insisted, therefore, that the starting point, not to be subject to doubt, could only be those things that are universally held. The example he gave for this, in 1949, was peculiarly unfortunate: that we would never walk on the Moon. This indeed illustrates how very difficult it is to get going if one tries to make the starting point explicit and unquestionably right. In the words of Søren Kierkegaard:

The method which begins by doubting in order to philosophize is just as suited to its purpose as making a soldier lie down in a heap in order to teach him to stand upright.

Starting point problems are by no means unique to science: for example, they figure importantly in linguistics in the definition and role of "heritage expressions." Presuppositions, starting points of the most fundamental kind, are, indeed, always only implied, but their truth is not dependent upon the truth or otherwise of the explicit proposition that they underlie. But since it cannot be got out into the open the presupposition's truth is not susceptible of verification except by making it explicit, in which case it will in turn have to become dependent upon a deeper presupposition and so ad infinitum. This is similar to the situation with logical systems where Gödel's result tells us that metasystems must be involved in an infinite regression if we foolishly embark upon the quest for ultimate self-consistency. As Stephen Hawking reported to us, the lady who thought that the world was carried on the back of a turtle, when asked what held up the turtle, could make her world picture self-consistent only by asserting that the turtle was held up by another turtle and that it was:

. . . turtles all the way down.

No!, the starting point to our sequence of presuppositions, of logical propositions, upon which our hoped-for analysis of the Universe depends, can, logically, only be psychological, in other words, we ourselves; the starting point is indeed rooted in our nature together with those inaccessible and irradicable prejudices. I do not believe that we may confidently assume that the Universe shares those same prejudices.

2. Our Intellectual Tools

The Move Toward Physical Science

Having now surveyed some of the considerations that must lie behind our detailed questioning of the Universe in terms of our humanity and the kinds of limitation that this must place upon us, I turn to more scientific matters as they are usually and narrowly defined: how do scientists go about their business; what is the nature of the intellectual machinery they employ? I suppose that most people will think of this as realism at last in place of the long chain of nonscientific considerations that stemmed from my earlier excursion into the nature of our thought processes and neurophysiological equipment for mounting them, starting with the other extreme of naive idealism. You are going to be disappointed: the realism of the hard physical scientist very soon, as you will see, slips into something quite closely akin to naive idealism and before long moves into realms fully as divorced from common sense as those that occasioned the profitless hurt to Dr Johnson's toe to which I earlier alluded.

The science of the physical Universe is a very rich and complex tapestry into which many diverse and wonderful scenes are woven. I certainly do not wish to consider any one part of the whole picture as any more important or intellectually demanding than any other: nuclear physics than biochemistry or astronomy than metallurgy, but the warp and woof of the ground upon which the picture is built up are the great underlying principles of quantum mechanics and general relativity.

Quantum mechanics deals chiefly, but not exclusively, with the very small while general relativity deals chiefly, but not exclusively, with the very large. At the moment these two great structures stand apart from one another, but most physicists now believe that they must be brought together and that in that bringing together we shall penetrate to the heart of the Universe. Of that bringing together I shall speak later but by way of introduction I shall give a brief and completely nontechnical account of each, stressing those aspects that connect most closely with our nature in relation to our own understanding. First, quantum mechanics.

Quantum Mechanics

I begin with one of the basic tenets of quantum mechanics, the technology usually thought of as submicroscopic although it is by no means necessarily always so, that we believe to underlie all the workings of the natural world. This basic tenet is that you cannot find out anything new about a system, large or small, without making an observation upon it, that is to say, without interacting with it in some way, and that this observation, this interaction, if it is to extract information from the system, must obviously and necessarily disturb the system under observation to some degree. The final state of the system, after the observation, will therefore not be the same as its initial state before the observation but will differ from it to a degree that cannot itself be known unless the observation was itself observed, which lands us in an infinite regression. So every act of the finding out of new knowledge about a system entrains a change in the system and also, notice, in the observer because if information is to be recorded by the observer, then the observer must also be changed thereby. It is obvious that the observer must be sufficiently complex to suffer and record that change.

I should remark, in passing, that the popular statement that all observations disturb the system observed is not strictly true. The exception, which I have carefully skirted around in the remarks I have just made, concerns the so-called von Neumann states where the observer, prior to making the observation, was already in the condition corresponding to the response to the state in which the system itself in fact already was. In that case the system is not disturbed by the observation and the observer is also unchanged, merely confirming, as it were, the guess made prior to the observation. But this type of observation does not change the observer's state of knowledge, does not bring in new information. But with that uninteresting, if important, exception we have the infrangible rule that observation changes the system observed and also changes the observer. We begin to see again the inescapable linkage between us and the Universe of our concern; in science, as well as in philosophy, the Universe is defined through our awareness of it.

The Wave Function. In quantum mechanics we describe the system whose behavior we are tying to discuss by a mathematical construct, the wave function, which tells us the probability, but only the probability, that our observation will reveal the system to be in a certain state at a certain time; most simply, for example, that an electron will be found at one place or another at a certain time. In general, the wave function will describe several possible states simultaneously. Only when we know for sure that the system is in one particular state will the wave function refer to just one state. In setting up the wave function that contains the information that enables us to make the prediction about the state in which the system might be found, about what future relative probabilities might be, we obviously use all the information we have available. That is all we can say about the wave function; it is a summary of our knowledge of the system and also of our ignorance.

After we have made our observation we have more information about the system than we had before because we now know for certain that it was in this condition rather than that, that the electron was here rather than there; probability has turned into certainty. Or so we naturally interpret it, although we have no real right to do so and, in any case, such retrospective information is not of much use: it is the future that counts. But the gaining of that information, the making of that observation, has disturbed the system to a degree that itself can be described only statistically so that the wave function that we now use for describing the future evolution of the system, for making useful predictions, must reflect that new uncertainty, as well as the new knowledge.

Do not mistake the situation: there is nothing random or capricious about the behavior of the wave function: if the system is not being disturbed, if no observation is being made upon it, the wave function behaves totally deterministically like perfect billiard balls on a perfect billiard table. But the information that evolves deterministically in this way relates only to the probability that a certain observation will bring a certain result: we cannot say with certainty what that result will be.

The Seat of Reality. Where does reality reside in the physicist's description of the natural world? Is it in the deterministically evolving wave function or in the system at the instant the observation is made upon it? If you choose the former, then we have almost the antithesis of Bishop Berkeley's idealism: the tree in the quadrangle exists only when it is *not* being observed. If you choose the latter, then reality is a poor, stuttering affair, real only from capriciously related instant to instant because continuous observation would mean continuous disturbance with chaotic, unpredictable consequences.

What should we make of this? Which choice should we make? In a significant sense it does not matter because we know how, unambiguously, to get out the answer to questions relating to actual observation and it works prodigiously well.

Practical Application. Consider, for example, the so-called *g*-factor of the electron, which is its magnetic moment divided by its angular momentum in appropriate quantum units, the latter usually just called its spin. Our simplest picture of an electron manifesting its magnetic moment by interacting with a magnetic field would be of the electron cruising along through space and time and latching directly onto the magnetic field. Relativistic quantum mechanics, as formulated by Paul Adrien Maurice Dirac in 1928, says that this would give *g* = 2: indeed *g* is very nearly equal to 2 but not quite; it is now known experimentally with very great precision thanks to work by Hans Dehmelt and his colleagues:

$$g/2 = 1.001159652193$$
$$\pm 0.000000000010$$

Quantum mechanics does not, however, say that all that happens when an electron manifests its magnetic moment is that it latches directly onto the magnetic field. It says that perhaps before that latching takes place, the electron emits a photon, a particle of light, the latching then takes place, and afterward the electron catches again the photon that it had previously emitted so that the latching does not take place with the simple electron but with the more complicated system of the electron with a photon as it were "in the air" at the same time, between being emitted and caught again. Alternatively, the electron might emit a photon as before, which, while in the air, breaks up into an electron pair consisting of an ordinary negatively charged electron plus its antiparticle, namely, a positively charged electron, a positron; one of the electrons of this midair pair then latches onto the magnetic field and then later recombines with the other, regenerating a photon, which is then absorbed by the original electron, which has therefore not itself directly latched onto the magnetic field at all.

Quantum mechanics says that to calculate the full *g*-factor we must take into account such processes as I have just illustrated, to all degrees of complexity; an infinite number of such forms of quantum electrodynamic midair gymnastics and, furthermore, that we must imagine that all these separately described midair activities are going on *simultaneously*. Picture a lady gymnast performing on a vaulting horse. Sometimes the gymnast simply runs past without a vault; sometimes she just dives over the horse without touching it; sometimes she does a handspring from it; sometimes she does a somersault, or two somersaults, sometimes forward and sometimes backward; sometimes she does one or two twists in the air. Now imagine the gymnast doing all these things *at the same time:* doing a forward somersault *at the same time* as doing a backward somersault *at the same time* as simply running past without doing anything: the whole Olympic performance *at the same time*. If you can imagine this you can begin to claim that you understand quantum mechanics, but not otherwise.

Richard Feynman who more than anyone brought quantum mechanics to its present-day point of practical perfection in terms of its detailed predictions wrote:

I think it is safe to say that nobody understands quantum mechanics. So . . . just relax and enjoy it.

And he should know. He also said:

Do not keep saying to yourself, if you can possible avoid it, "But how can it be like that?" because you will get down the drain into a blind alley from which nobody has yet escaped. Nobody knows how it can be like that.

Let us not follow those lost souls down their drain.

Although we cannot imagine all that quantum mechanics seems to bid us to imagine, we can do the calculations that quantum mechanics bids us to do. In the case of the electron's *g*-factor, to the immensely elaborated degree to which it has been done, largely by Toichiro Kinoshita, we predict:

$$g/2 = 1.001159652190$$
$$\pm 0.000000000110$$

where the uncertainty is due almost entirely to the experimental uncertainties in the value of the combination of the natural constants (the so-called fine structure constant) that enters the theoretical expressions, with only negligible contribution from uncertainties associated with the theoretical expressions themselves. The agreement between theory and experiment is staggering. Surely the truth of the procedure cannot be denied although the meaning of that truth in the sense of our grasp of any underlying reality is negligible, if that.

Landau's Ghost. But now I must even withdraw that "surely" and retract my earlier reference to our adding together all the infinity of possible midair gymnastics. It can be shown that if the calculation is carried far enough, to a degree that is totally impracticable but that can be talked about in principle, beyond the point where our gymnast makes more than a hundred somersaults in midair between takeoff and landing, we shall not continue to get closer and closer to the right answer but will begin, very very slowly, to move away from it and diverge into nonsense. This is known in the trade as Landau's Ghost. It probably tells us that our brilliant successes with quantum electrodynamics, such as g-2, the renormalization program of Feynman, Schwinger, and Tomonaga, only quantify an approximate viewpoint, an extremely good approximation but only an approximation nevertheless, and that we must look for a deeper starting point that will not involve our recourse to an imagined infinity of midair acrobatics on the part of our electron and will thereby lay Landau's Ghost.

Such was the frequently and clearly expressed view of Dirac himself when, for example, he said in 1980:

... something has to be changed. ... We need some new mathematics ... I believe that the only true answer will be obtained when someone is able to think of this new mathematics. . . . In spite of the (renormalization programme's) successes we should be prepared to abandon it completely and look on all the successes . . . as just accidents when they give the right answers.

And when, much earlier, he had ended his famous textbook on quantum mechanics with:

It seems that some essentially new physical ideas are here needed.

Dirac always insisted that one should give the mathematics its head even though one did not, for the time being, see the sense of where it was leading one—he used to say that his own famous equation was more intelligent than its author. He wrote:

. . . one should allow onseself to be led in the direction which the mathematics suggests . . . one must follow up this mathematical idea and see what its consequences are, even though one gets led into a domain which is completely foreign to what one started with.

Dirac again:

The most powerful method of advance . . . is to employ all the resources of pure mathematics in attempts to perfect and generalize the mathematical formalism that forms the existing basis of theoretical physics and after each success in this direction, to try to interpret the new mathematical features in terms of physical entities.

Dirac again:

People at the present time are making the mistake of continually trying to develop the physical ideas which they have gotten used to. . . . What one really needs is a new kind of mathematics . . . one should not stick to the standard ideas and just try to push on with them.

We must wait and see. To date we know of no phenomenon that, of itself, demands such a radical rethinking although the sensational and mysterious phenomenon of positron-electron pairs found in heavy-ion bombardments at GSI, Darmstadt, is thought by some to signal a new phase of quantum electrodynamics, so-called strong quantum electrodynamics, that bypasses the gymnastics and may lay Landau's Ghost.

The EPR Paradox: Nonlocality. We leave the triumphant successes of quantum mechanics, albeit slightly uneasily, with no more idea about the seat of reality. If we believe that the underlying wave function, whose deterministic evolution enables us to make all our highly successful and very practical predictions throughout the physical world, is itself in some sense "real," then we must put it outside the scope of the physics that it itself represents.

This is most dramatically seen in the famous Einstein-Podolsky-Rosen paradox, the EPR paradox, which arose from Einstein's grave puzzlement about the nature of reality in quantum mechanics and which bids us to consider the wave function of a system evolving in space and time. The wave function spreads out more and more until an observation is made that pins down that aspect of the whole wave function that relates to that observation, say, one particle out of several that the overall system, the overall wave function, may comprise. That pinning down brings new information and since the wave function is a statement of our total knowledge of the system as a whole, the wave function of the system as a whole must instantaneously change.

As a simple example, consider a system that consists of a particle that at the word "go" explodes into two other particles, identical to each other, that shoot off in opposite directions, and that is all we know. Then prior to any observation the wave function describing the system must be spread out uniformly in all directions because we do not know where the two product particles have gone. But as soon as we find one particle, say, due north of the point of explosion, we know that the other particle must have gone off due south, so the wave function describing that second particle, instead of being spread out in all directions as it was before, must collapse instantaneously into a form permitting the second particle to be found only due south. The collapse is instantaneous because our knowledge undergoes an instantaneous augmentation and the wave function merely represents our knowledge. But the first particle may have been observed a million miles away to the north so that the second particle would now be a million miles away to the south. The wave function describing the second particle must collapse instantaneously although to send an actual message to the second particle from the first by the fastest means available, namely, a light signal, would take more than ten seconds, in the course of which time we may, in fact, have observed it in its expected, newly precise, position: the wave function must collapse before we can possibly reach it with a signal telling it to do so. So if we wish to say that the wave function that collapses has a reality, then that reality does not conform to the requirements of relativity, which themselves control its deterministic evolution in the absence of new knowledge.

This has been a very crude exposition of the EPR paradox; its full development shows how not just the crude fact of the second particle's position but also delicate matters relating to its state of internal organization, such as its spin or polarization, appear to be transmitted instantaneously following observation of the first particle rather than wait on that usual limiting speed, the lumbering velocity of light.

All this is not hypothetical fancy: experiments have been carried out, most recently by Alain Aspect in the 1980s but going back to Bleuler and Bradt and to R. C. Hanna in the 1940s, that appear to confirm this instantaneous communication between two particles far apart on the macroscopic scale. Of course, such experiments confirm instantaneous communication only if you believe in the reality of the wave function, which must, therefore, adjust itself in the pseudomaterial sense as observations are made.

The Copenhagen Interpretation. Physicists who, in this sense, believe in the reality of the wave function are therefore morally obliged to admit their belief in psychokinesis. If, on the other hand, you do not worry about the wave function's being the seat of reality and regard it, as on the Copenhagen interpretation of Niels Bohr, as a deterministically evolving mathematical representation of our incremental state of knowledge, and nothing more, then there is no problem: the initial particle explodes, you set up the wave function describing the totality of the system, as a single linked system of the two product particles; you sit back and enjoy the evolving show—and get the right answers.

I greatly prefer that second viewpoint, even though it reduces reality to a series of snapshots as observations are made, and I therefore reject the common but facile answer to the question "Is the electron a wave or is it a particle?" namely, "Sometimes one and sometimes the other." The nearest I will go is to paraphrase a remark that E. J. Williams made fifty years ago and say:

"The electron is, of course, a particle, the wave is in the mathematics."

In operational fact, if the electron actually does something, it looks like a particle, but if you are not looking at it, your knowledge of where it might do something when it actually shows up, but not, in detail, what it might actually then do, has to be represented by an evolving wave function that has no other sense or purpose than to encapsulate that knowledge and ignorance.

The apparent ability of disconnected parts of a system to communicate with each other instantaneously is what worried Einstein in framing the EPR paradox; he called it "spooky actions at a distance." It is now known in the trade as nonlocality and is at the heart of the formalism of quantum mechanics; it can no more safely be dug out and thrown away than your or my heart can. Quantum mechanics says that a system remains a single quantum mechanical entity, subject to the overall deterministic evolution of its overall wave function, even though its constituent parts may have lost contact with each other and have no causal mechanism for comparing notes and refreshing each other's memories of their common origin and early life together. Now there is obviously no objection, from Einstein or anybody else, to those now-separated parts of the system retaining definite relationships with each other that were established at the time when they were in causal contact; that would be understandable memory. But nonlocality means that relationships between the separated parts of the system appear to be established instantaneously following an observation upon one of those parts and in a manner that is not due to understandable memory. That is indeed spooky, but it is quantum mechanics and it works. Please do not distress yourself; in N. David Mermin's summary of the Copenhagen interpretation: "Shut up and calculate."

The Transactional Interpretation. There have been many attempts to reinterpret quantum mechanics so as to expunge the spookiness while retaining the successes that quantum mechanics enjoys. Such attempts usually turn out to be equivalent to the Copenhagen interpretation; the spooks return. A current attempt by John Cramer invokes the time symmetry of all fundamental physical equations, which leads to the mathematical appearance of "advanced" solutions (in which time "runs backward"), as well as "retarded" solutions in which time runs the usual way. "Retarded" means that the effect follows the cause; "advanced" means that the effect precedes the cause; because of this, the advanced solutions are usually simply thrown away because they seem to be nonphysical and obviously violate causality; however, they cannot be thrown away on purely mathematical grounds.

It is, however, possible, by the use of suitable boundary conditions, to preserve causality while retaining both sets of solutions, as was done by Wheeler and Feynman for ordinary classical electrodynamics in 1945, leading to exactly the same results as the conventional retarded-waves-only method. Cramer has now developed an analogous approach to quantum mechanics that he calls the "transactional interpretation," seeing it as a "handshake" across space-time, a

. . . two-way contract between the future and the past for the purpose of transferring energy, momentum, etc, while preserving all of the conservation laws and quantization conditions imposed at the . . . boundaries of the transaction.

We shall have to wait and see whether this idea is helpful in a major way, but it does bring out the fact that the symmetry of the fundamental laws of physics with respect to the arrow of time is in conflict with our own human perceptions of the matter; I shall have another word about this later.

Hidden Variables. All this is to say that we know how to get the right answers out of quantum mechanics although we do not know what the process of gaining those answers itself means. This should not be surprising since quantum mechanics deals with matters remote in scales of space and time from our household experiences on whose basis we have, by definition, set up our common sense. But this is not to say that there are not continuing searches under way for a reality in, or behind, the wave function, perhaps a deeper determinism underlying the apparently purely probabilistic interpretation of the wave function, namely, the probability, and no more, that something will be found at a particular place at a particular time in a particular condition.

The irregular and flickering flight of a bat at night, apparently random and purposeless is, in fact, deterministically guided by the desperate twists and turns of the escaping bug that it is trying to catch. Possible deeper layer deterministic pictures of the quantum world go by the trade name of "hidden variables"; their possible role, and tests for their possible existence, are restricted by the von Neumann relationships, now sharply modernized into Bell's Inequalities.

It is interesting and salutary to hear John Stewart Bell, the author of much of our current critical reexamination of the bases and meaning of quantum mechanics, speaking in 1986:

... progress is made in spite of the fundamental obscurity of quantum mechanics. Our theorists stride through that obscurity unimpeded ... sleepwalking? The progress so made is immensely impressive. If it is made by sleepwalkers, is it wise to shout "wake up"? I am not sure that it is. So I shall speak now in a very low voice.

Let us not, then, strive too officiously to understand quantum mechanics but let us acknowledge that it is the basis, our only basis, for treating with the submicroscopic world and for any aspect of the macroscopic world for which its considerations have relevance, including, indeed, as we shall see, the Universe as a whole.

Schrödinger's Cat. On the way to the Universe as a whole it is instructive to stop off at intermediate points. Perhaps the most famous of these is Schrödinger's Cat. Schrödinger bade us consider a cat inside a black box that contained a lethal device that had a 50-50 chance of going off before we opened the box to have a look. Just before we open the box, all that we know, therefore, is that the chances are 50-50 that when we look inside we shall find a live cat or a dead cat. We must then describe the cat at the moment before the box is opened by a wave function that expresses this uncertainty: the wave function must have equal amounts of dead cat and live cat. If we believe that the wave function is itself reality, this means that the cat is half dead and half alive before we open the box and then instantaneously becomes one or the other as we look inside and the wave function collapses into the one or other state.

If you are a strict Copenhagenist you have no problem because the wave function is not itself the cat but merely tells us all that we know about it. I have oversimplified the problem and its resolution: a full discussion takes us into questions of the completeness of the wave function as a description of all relevant aspects of the system, including matters to do with its essential environment and so on. But Schrödinger's Cat at least enables me to point out that quantum mechanics must apply to all systems, big as well as small. A current example of this is its application to a development of superconducting SQUID magnetometers in which we speak of an object some millimeters across existing in a giant quantum state, or, more particularly, in a superposition of such states. This can indeed be discussed as a kind of properly defined Schrödinger's Cat.

Large Systems. Another stopping-off point concerns, not the quantum description of large and complex systems such as SQUIDS and cats, but rather of large and simple systems such as the hydrogen atom in a state of very high excitation with its electron therefore at a large distance, say a meter, from its proton. This is then a macroscopic object such as we are used to handling, literally, in our everyday lives, and it is interesting to ask what quantum mechanics has to say about it. We should expect to be able, if our vision were sufficiently acute, to see the electron going round and round the proton in a circle. But does quantum mechanics not tell us that we, in fact, cannot do this? Have I not explained that every observation disturbs the system so that if I try to watch the electron going round and round the proton, the observations through which I do this must necessarily knock it off its orbit? And have I not explained that the electron does not, in fact, have an orbit at all but is represented by a wave spread out in space?

We certainly do not have this kind of anxiety in our everyday macroscopic world: as the baseball soars toward the outfielder the outfielder does not feel that he should not look at it lest his observations knock it off its trajectory and make him miss the catch. Why is the hydrogen atom in its macroscopic state any different? The answer is that it is not different when we ask a question whose answer has a macroscopic visualizability. Thus when we calculate the wave function for the hydrogen atom in a highly excited state a meter across with its electron going in a circle, we find that the electron's wave function, although it indeed extends throughout all space, is effectively confined to an ordinary classical-looking orbit, almost all the probability density that it represents being found within a very few millimeters of the circle of radius one meter, and that we may make observations to locate the electron on that orbit without displacing it significantly from it. When the atom is big enough the electron can behave just like the baseball. In short, when we use quantum mechanics for describing things on our own scale, even though those things are of an atomic nature, the answers we get are those of common sense; this is somewhat comforting.

Is the Wave Function All? Although, as we have seen, the wave function is not in any objective sense itself reality, it represents that reality, if such there be, and is our mechanism of access to reality insofar as it tells us as much as it is possible to tell about what is going to be the result of observations. If we are ever going to give an account of "everything" it will have to be through, or consonant with, the quantum theory. Or will it? The qualification is what I have already discussed, namely, the possibility, to me a very real one, that the Universe cannot be described just in terms of mournful numbers and that things are, indeed, not what they seem, although I do not believe that that is what Longfellow had in mind. But, begging this reservation, I am now exposing the conventional view of the card-carrying physicist and shall continue with it.

Consciousness and the Quantum World. I have repeatedly stressed the role of observation, and hence of the observer, in our view of the quantum world. Although the system of which we speak, whatever it is, evolves deterministically when left strictly to itself, it can only be brought out into the open, one might almost say be brought into being, certainly be brought into established significance in the sense of trading with a recording system, by the act of observation. And since we can give no meaning to observation other than by some eventual lodgment in our consciousness we may seem driven to the conclusion that a Universe without consciousness simply cannot be discussed. Consciousness, expressed through our logical and linguistic machinery, is an essential factor in our discussion of the natural world. And, since that consciousness rests upon those disturbing observations, consciousness must be a factor in the development of the natural world.

Such views have been put forward, for example, by John Archibald Wheeler and by Eugene Paul Wigner and must command high respect, although the inference may seem strange, namely, that those parts of the Universe harboring consciousness must develop differently from those that do not. Wheeler's way of expressing it in 1988 is that:

The world is a self-synthesizing system of existences. . . . That system of shared experience which we call the world is viewed as building itself out of elementary quantum phenomena, elementary acts of observer-participancy. In other words, the questions that the participants put—and the answers that they get—by their observing devices, plus their communication of their findings, take part in creating the impressions which we call the system: that whole great system which to a superficial look is time and space, particles and fields. That system in turn gives birth to the observer-participants.

In other words, to paraphrase but not to caricature Wheeler, he is saying that we literally create the Universe by our formative participation in its development. As Cesare Pavese says:

To know the world one must construct it:

Or as Erich Fromm puts it:

Man unites himself with the world in the process of creation.

If one wishes to escape this conundrum of the intimacy of the relationship between the "material" Universe and consciousness, to whatever degree of intimacy one may find oneself to be led, and it would indeed be difficult to be led further than Wheeler, whose position may be seen as a logical conclusion of the exercise, one may take refuge with the extreme behaviorists, thereby disembarrassing oneself of what James Broadus Watson, already in 1915, had referred to as "the rubbish called consciousness." But personally I hanker, perhaps sentimentally, after the idea that there is some sort of "me" that is conscious and that sees, hears, and enjoys.

Wheeler's participatory Universe recalls the remark of Angelus Silesius in the middle of the seventeenth century:

He who would rightly comprehend the world must be now Democritus, and now Heraclitus.

(It was Democritus who asserted that the world is constructed out of the material entities that we now call atoms, and Heraclitus asserted that change itself is the basis of knowledge—change in his case as among the four classical elements, in Wheeler's case in the sense of information transfer.)

States of Mind. Observations depend upon interactions, and one may regard those interactions as equivalent to observations, provided that, as I remarked before, the interacting-observing system is sufficiently rich in its own structure to define and then to record the various possible states of concern of the system with which interaction takes place. We can speak meaningfully only of those states of the observed system that correspond to states of the observing system, and the observation simply sets up a correspondence between those two sets of states. Before the interaction took place, the system to be observed was described by a wave function quantifying the possibility that a future observation would reveal it as then, after the observation, being in one or other of a number, perhaps small, perhaps large, of alternative states. This quantification is not retrospective: it does not assert that the system, before the observation, *was* then in one or other of the range of possible states; such a statement would have no sense because it could not be verified; you cannot talk about the system being in a definite state prior to the observation that determines which state it is then in. The observation, the interaction, singles out one fom the many possible states of the observed system and simultaneously switches the observing system into whichever of its recording conditions it is that corresponds to the state observed; both observed and observer are affected. So after the observation the observed system is in a definite state, which then evolves deterministically on its own until we look at it again. But what has happened to all those other possible states represented with various probabilities in the wave function before the lucky one was picked out by the observation?

If one follows the Copenhagen line that I discussed in the context of the EPR paradox, one says that those other states have simply been dismissed; the wave function has collapsed into the single state indicated by the observation; prior to the observation all the states were only states of mind; now only the observed state has reality, and only now, not then: at that earlier time it too was only a state of mind.

Many Worlds. There is another viewpoint, one that links the wave function more closely with reality but that does not involve it in the sudden and convulsive changes following its redefinition through observation that we saw seemed to prevent our saying that it was more than just a state of mind. This other viewpoint, put forward by Hugh Everett in 1957, is the Many Worlds interpretation of quantum mechanics. This viewpoint says that all those possible states of the system, simultaneously represented in its wave function, *are* all real and furthermore *continue to be real* and manifest after the observation that singles out one of them as of our special concern. That special state is the one that we, the observing system, have in our consciousness, but the others remain and go their own way, accompanied by the observing system, that is to say ourselves, in its own appropriate corresponding state, in a different mode of consciousness.

These other states, each accompanied by its own version of ourselves, go their own way, undergoing their own subsequent interactions in their own worlds, each interaction producing yet other branchings, yet other splittings into an increasing myriad of other worlds inhabited by a myriad other aspects of ourselves. These other worlds are, of course, disconnected from our own, and those other aspects of ourselves are similarly disconnected from the "me" and "you" that are writing and reading this book. We run a kind of slalom course through all possible worlds: at each slalom gate we could have branched off onto some other slalom course, could have gone off some other way into some other world, and, indeed, so we did: it is just of this choice, this track, this world that this aspect of ourselves is conscious; those other aspects of ourselves are enjoying other worlds with which we have no contact.

But all those other worlds, all those other slalom courses, all those other Universes, are just as real as "ours" and all those other aspects of ourselves that are running those other courses are just as real as the "we" of our present consciousness. It is a frenzy of multiple schizophrenia. As Bryce DeWitt, one of the deep analysts of the quantum world, puts it:

The Universe is constantly splitting into a stupendous number of branches, all resulting from the measurement-like interactions between its myriads of components. Moreover, every quantum transition taking place on every star, in every galaxy, in every remote corner of the Universe is splitting our local world on Earth in myriads of copies of itself. . . . The idea of 10^{100+} slightly imperfect copies of oneself all constantly splitting into further copies which ultimately become unrecognizable is not easy to reconcile with common sense. . . .

To the last of those sentiments I feel that we may readily subscribe.

As to the splitting of ourselves, we cannot avoid this on the Many Worlds view; the only question is the conditions under which this comes about and the relationship between our splitting and that of the rest of the Universe. Frank Tipler expresses it like this:

A human being can properly be said to split only if he undergoes an interaction with the rest of the Universe which causes a change of state in his memory, in which different states of his memory are correlated after the interaction with different states of that portion of the rest of the Universe with which he interacted.

This moderates the frenzy of the schizophrenia but not, you may think, much. It is indeed difficult to bring in common sense at this stage, but Einstein went partway when he rebelled against the infinite multiplicity of Universes that this Many Worlds picture requires: "Surely," he said to Everett, "you cannot mean that a mouse looking at the Universe splits the Universe into those myriads of states"; Everett, faced with so formidable an interlocutor, conceded that it was perhaps not the Universe but rather the mouse that was split.

All this is perhaps not as modern and novel as you may think. One hundred years ago Poul Martin Moeller was writing:[‘]

I get to think about my own thoughts ... I even think that I think of it and divide myself into an infinite number of "I"s who consider each other. ...

Although this Many Worlds view may seem fanciful, even absurd, it is the only way in which we can consistently give an element of reality to the wave function that underlies everything that we know about the submicroscopic world and that we might therefore hope to be real if anything is; in particular it avoids the patently unphysical collapse of the wave function following an observational interaction and gives to reality a continuity denied it by the Copenhagen view; in this sense it must be powerfully appealing to us. It is logically self-consistent and, like naive idealism, cannot be refuted. It gives the same answers as the conventional Copenhagen approach for small-scale problems such as laboratory experiments, but it sits much more easily than the Copenhagen view with the grand cosmological problems of the beginning and early history of the Universe to which I shall shortly turn. Despite its disjunction with common sense, however, most cosmologists and the particle physicists with whom they increasingly consort now espouse this Many Worlds view with its staggering implications because it is the only quantum language so far devised through which it may be possible to construct a physically sensible wave function of the Universe in toto. To this I shall return.

Human Beings in the Universe. If we describe the Universe as a whole, then the observer is manifestly part of that description and subject to the same quantum laws as the rest; the observer cannot stand back from the Universe; any move, any observation the observer makes must affect the rest of the Universe because the observer and it are linked through the overall wave function of which they are both part and that must simultaneously describe them both; and, similarly, any change in, any evolution of, the observed part of the Universe must entrain a change in the observer. We may well echo with Omar Khayyam's "impatient one":

Who is the potter, pray, and who the pot?

From there it is only a small step to observe that if, as is indeed the case, we are ourselves part of that Universe that we are describing, then that description will itself ultimately be determined by the laws under which the Universe operates, so that our conclusion about the nature of those laws, and so about the nature of the Universe, will itself be determined by those same laws. It seems as though we have no choice in the conclusions we shall draw about the nature of the Universe, but there is nothing in this argument to suggest that those conclusions must be the correct ones. Perhaps my repeated insistence that there may be something more to the

Universe than just numbers may enable us to escape this unpalatable conclusion.

I will now leave quantum mechanics, which, while being by no manner of means Earthbound, as we shall soon see, is chiefly manifested through the very small, and look outward toward the very large, our Universe and the broader cosmos of which, as I shall hold, it is only a part.

Cosmologies

We have always gazed at the stars with wonder. We have grouped them into constellations to which we have attributed powers of destiny from which we have only partially broken away:

> The fault, dear Brutus, is not in our stars,
> But in ourselves, that we are underlings.

And within and beyond the stars and the planets we have constructed cosmologies that usually mirrored our own social or ethical structures.

The cosmologies of the classical Greeks are well known and I will not dwell upon them. But the underlying motivations for those cosmologies are not so well known, and it is useful to glance at them because cosmologies are a product of the *Zeitgeist,* as must be the case with ours today also, and it is salutary to recognize this.

The classical touchstone was perfection, of which there had to be a measure. In "De caelo" Aristotle linked degree of perfection to distance from the Earth, which was later understood as meaning nearer to God. The author of "De mundo" writes:

> . . . the Earth and things upon the Earth, being furthest removed from the benefit which proceeds from God, [this within an Earth-centred cosmology of course] seem feeble and incoherent and full of much confusion.

In the second book of "De caelo" Aristotle said that it was rather the variety of motions carried out by a celestial body that measured its nobility or degree of perfection, the fewer the better. This made the fixed stars the most noble and the planets less so because of their to-and-fro motions among the stars. But then the Sun and Moon were also noble because they had simple motions. The Sun was evidently the noblest planet, but why, then, was it placed fourth rather than on the outside? The explanation was in another form of nobility that it was evidently displaying: to diffuse its beneficence to best advantage; if it had been on the outside we should have frozen and if on the inside we should have burned.

So it continued until the decline of scholasticism in the seventeenth century when Copernicus, Galileo, and Newton, not without some difficulty, took over. But although we have moved away from criteria such as nobility as a measure of a scientific theory, we still place great store by the sort of perfection, or echo of perfection, that we find in symmetry schemes, broken or otherwise.

Some cosmologies, such as those of the Indian subcontinent, mirror local social structures in their extraordinary complexity and hierarchical arrangements. They also tend to be accretive, never rejecting but always adding new features, even though this generates evident inconsistencies. And they sometimes achieved extraordinary precision in their own fantasy land. Thus in the cosmology of the Jains, astronomical distances were measured in rajjus, which is the space covered by God in six months flying at 2,057,152 yojanas per blink. Now a yojana is defined as 1/1600 of the Earth's diameter and a blink is 1/5 second, so a rajju comes out at 137 light years. This is a remarkably appropriate unit in which to measure astronomical distances and is also the most convincing derivation I know of the reciprocal of the fine structure constant although it is pressed hard by Victor Weisskopf's revelation that 137 is the number associated in Jewish mysticism with the word *cabala*—what else? In science we do not model theories on social structures any more than on concepts of nobility, although the notion of hierarchy remains one of central importance.

Now, with renewed interest in anthropomorphic influences, I

turn back to the science of today as it applies, at least in the first instance, to the largest scales of physical size.

Large-scale Structure. The large-scale structure of our world is dominated by gravity. Certain dramatic cosmic manifestations are primarily electromagnetic in origin; indeed, the largest known spatially ordered single objects, the "bow tie" radio galaxies, which range up to more than a hundred times the size of a large spiral galaxy, have such an origin, but they and related objects have almost no mass, cannot properly be regarded as structures in the sense of depending upon the interactions of their constituent parts, and have no significant interaction with other bodies. Gravity, as described by Einstein's general theory of relativity, is the only long-range mechanism for the communication of order in the Universe as a whole.

Our Place. General relativity is the theory of the very large. Small and large are relative terms and in using them I place human beings in the middle as the measure from which we move to the one or to the other. The smallest distance to which our experiments have so far probed is some 10^{22} times smaller than a human being; the largest organized gravitating objects to which our astronomical observations relate, namely, the spiral galaxies, are some 10^{21} times larger than a human being. The smallest distances that enter into our theoretical descriptions of the physical Universe are some 10^{35} times smaller than a human being; the edge of the visible Universe is some 10^{26} times as distant as the size of a human being. It is impossible to gain the slightest personal feeling for the meaning of numbers such as these: the price of a piece of gold the size of the Earth is about $\$10^{29}$ (at today's prices I cautiously add).

We are in the middle and, furthermore, utterly remote in terms of our own experience from, in opposite directions, the smallest and the largest objects that are of our concern when we try to understand the Universe. But it is only from our own household experiences that we have derived our language and our capacity for talking about and understanding the Universe and all that it contains. Small wonder then if that language lets us down in those remote regions of the very small and very large. The wonder is rather that we can make any kind of sense of it at all. As Einstein said, the only incomprehensible thing about the Universe is that it is comprehensible. Or, perhaps, as I would put it, that we can force it into our comprehension.

General Relativity. General relativity deals with gravity, the force that, as I have remarked, is responsible for the large-scale structure of the Universe, the clustering of matter into stars and galaxies, the motion of the planets around the Sun, and the sticking of ourselves onto the Earth. But I have already sold the pass by speaking of gravity as a force: in classical, that is to say, non-quantum-mechanical, terms it is not; it only seems to be.

It goes like this: the other three so-far-recognized forces that, together with gravity, shape, or very largely shape, our natural world as we now see it, namely, the electromagnetic force, the strong nuclear force, and the weak force of beta radioactivity, have their being within a preordained space, which is usually pictured as a kind of rectilinear coordinate system, and manifest themselves through the exchange of messenger particles between the entities acted on by the forces: for example, the electric force is due to the exchange of photons between electrical charges; the strong force is due to the exchange of gluons between quarks; and the weak force is due to the exchange of the W and Z particles between quarks or between leptons, which are electrons and neutrinos, or between quarks and leptons. These three forces are all quantum mechanical forces and are described in the same coherent language. It is thought that they are, indeed, in a sense, alternative aspects of what would appear as just a single force at sufficiently high energies or temperatures but that has split into three apparently unrelated forces in our own low-energy, low-temperature world. More of this later, but for the moment let us regard the three quantum forces as being in contradistinction to gravity, which is not yet unambiguously understood through the language of quantum mechanics and therefore, in the way in which we use it in practice, is spoken of as a classical force.

Curved Space. Einstein, in his general theory of relativity, taught us that gravity does not operate in the same sort of rectilinear preordained space as the quantum forces but that a (large) gravitating mass deforms the space around itself and turns the straight lines of the rectilinear coordinate system into curves along which another, small mass in that neighborhood will tend to run. The small mass does the best it can to go in a straight line, but since the straight lines have become curves, it goes in a curve. Looked at from the outside the path of the small mass curves toward the large mass so we say that an attractive force is operating between them, but this is an illusion. A useful analogy is of a rubber sheet stretched out horizontally. With no masses around, the sheet is stretched out flat, but when a heavy mass is placed upon it, the sheet deforms and what previously were straight lines on it become curves. A small ball now placed upon the sheet will run down toward the heavy mass as though the heavy mass were exerting a force directly upon it, whereas the motion of the small ball is really due to the curvature of the sheet by the heavy mass. In general relativity space is, indeed, defined only through the masses that populate it and give it its structure; space without matter does not exist; matter defines and curves space.

The general theory of relativity is complicated but beautiful and aesthetically extremely satisfying. In the limit of comparatively weak gravitational fields, such as are experienced by planets and spacecraft, general relativity reduces very nearly to the familiar inverse square law of Newton, although even there we find small departures from Newton's previsions that are accurately accounted for by Einstein. Many such small departures have now been measured accurately, thanks to the technologies of space and radar, and in every case general relativity has been confirmed.

More dramatic confirmation comes from the experimentally strong inference that certain astronomical systems in rapid rotation can radiate some of their energy away through the emission of gravitational waves and from the observation that light from a distant astronomical source can be split if there is a gravitating mass almost on the line of sight from the Earth so that we see a multiple image of the source rather than a simple one. We know nothing contrary to general relativity, and although it cannot be said to be fully proven, it is almost universally accepted as the simplest self-consistent language in which to talk about gravity.

Therefore, until we know better we use general relativity as given, even in circumstances where it is strong in the technical sense of implying behavior radically different qualitatively from the previsions of Newtonian gravitation. But for the moment, it remains a classical theory.

Most people feel that gravity should be brought into the quantum fold, with or without a full integration with the three quantum forces of which I have spoken, so that we can at least speak about gravity and the other forces in the same language; at the moment we cannot do this and, in terms of its language, general relativity is simply inconsistent with the other forces: if we try to talk of classical general relativity in quantum language we generate nonsense. To this I shall return shortly. But there are others who believe that, because of its very different treatment of space, general relativity cannot itself be turned into a "normal" quantum theory such as can be wedded to the present three quantum forces, although the results of its actions must in any case be subject to overall quantum restrictions. As Wolfgang Pauli said:

What God hath put asunder let no man join.

But this view is increasingly unfashionable.

Black Holes. So we take general relativity literally and let it rip. When we do this we quickly find that there are indeed circumstances under which the behavior of matter is utterly different from our experience even if we broaden that to include stars and galaxies. This is the realm of the black hole.

I have mentioned that light is acted upon by gravity and that this has been checked out quantitatively against general relativity's predictions. But if you press this far enough, general relativity says that if the gravitating object is sufficiently massive and sufficiently dense, then it curves space so severely that light originating within it cannot get out at all, and if light cannot get out, then it is certain that nothing else can. Light can get to a certain distance but then has to turn back at what is called the event horizon. This is the black hole. Of course, things can fall into the black hole: so much the worse for them.

The mere passing of the event horizon does no harm, but once in you cannot get out or even send for help. It is even worse than that: general relativity requires, as first demonstrated by Roger Penrose in 1965, that at the center of the black hole there must be, literally, a point at which all the matter out of which the black hole has been made is concentrated; since the point has zero volume the density must be infinite, the curvature of space must be infinite, and everything breaks down; even general relativity, which has got us into the pickle, cannot make any predictions; the laws of Nature, together with space and time themselves, have disappeared. This is called a singularity and into it anything passing within the event horizon, astronauts and all, must totally disappear without trace or hope.

The only things that are not totally obliterated within a black hole are mass, angular momentum, and electrical charge; in particular the black hole continues to exert gravitational force upon bodies outside its event horizon and can suck them in. The event horizon itself can be quite large: that corresponding to a black hole mass equal to that of the Sun is about 3 kilometers in radius while that corresponding to a mass equal to that of an average galaxy—say, of 10^{11} solar masses—is about 10^{10} kilometers in radius. Inside the event horizon of a black hole nothing much happens to begin with, but you are drawn toward the singularity and eventually torn to pieces and reduced to nothing by the most beautiful mathematics imaginable. For all we know we are ourselves living at this moment inside a gigantic black hole; it is perhaps a comfort to know that there is no point in trying to do anything about it.

Evaporation of Black Holes. All that I have just said must now be qualified in an interesting way. I have said that not even light can get out of a black hole, but I have also said that the gravitational force penetrates the event horizon. And I also recounted earlier how pairs of particles and antiparticles, such as the negative and positive electrons of the quantum mechanical gymnastics that I described in connection with the magnetic moment of the electron, can arise spontaneously in space as fleeting quantum fluctuations. If they are isolated in space, then the particle and antiparticle must indeed get back together again with nothing to show for their brief attempt at expressing their individuality. But if the pair emerges in a field of force, such as the gravitational field of the black hole, then the particles can interact with that field of force and their subsequent history can be affected by it. In particular, in our present case, one member of the pair might fall into the black hole and the other be liberated into outer space taking with it some of the energy that previously was represented by the mass of the black hole. In other words, we here have a mechanism for getting energy out of a black hole, despite all appearances. But it is, note, a quantum mechanical mechanism, not a classical one, even though we continue to describe the gravitational effect itself in classical terms.

This wonderful prediction of the effective evaporation of black holes was made by Stephen Hawking in 1973, who showed that the rate of evaporation was exactly what would be expected from the ordinary laws of thermodynamics; in other words, a black hole may be characterized by a temperature. Of course, it is of little comfort to the astronaut who has fallen into the black hole because the astronaut comes out in the end as elementary particles. And the end can be a long time coming: a black hole of the mass of the Sun takes about 10^{66} years to evaporate, whereas one of the mass of a galaxy takes about 10^{100} years. On the other hand, a very light black hole, if it could be made, might evaporate very quickly: one of mass 10 micrograms, to which I shall refer shortly, would last only about 10^{-43} seconds. As an evaporating black hole gets lighter and lighter its evaporation gets quicker and quicker; the end is explosive in its rapidity and would be a most dramatic event.

So black holes are not forever, although those of the size that might now be being made, of the mass of a star and upward, last for a very long time; when they evaporate they release elementary particles that are not of immediate use in putting back together the life that may have fallen into the hole over the years; nor are the released particles released in such a form and in such a manner that it is imaginable that life-supporting circumstances could ever arise in the future.

White Dwarfs and Neutron Stars. Although black holes may seem to be very fanciful objects, it appears inevitable that they should form not all that infrequently as the end product of stellar evolution. Stars shine by converting the elements they contain into heavier elements by ordinary nuclear reactions that release energy. But these reactions must come to an end when they reach those elements that are most stable, around iron, beyond which nuclear reactions tend to cost energy rather than provide it.

At this stage we have a competition between the force of gravity, which is tending to make the star smaller, and the resistance to compression of the electrons that the star contains, which provides a kind of spring against which gravity works. In 1928 Subrahmanyan Chandrasekhar showed that if a star were less massive than about one-and-a-half Suns and had used up all its fuel in this way, the balance between gravity and the electron spring could be struck and the star would settle down into an object with a radius of only a few thousand miles and therefore a density of tons per cubic inch; these stars are called "white dwarfs" and many are known.

If the star were more massive than this Chandrasekhar limit, as it is called, then the electron spring would not be powerful enough to resist the squeeze of gravity and the star would shrink further. Eventually the electrons find continuing resistance futile and disappear into the protons, converting these into neutrons and emitting neutrinos, which escape from the star so that the whole star becomes one gigantic nucleus of neutrons: a neutron star of radius of only a few miles and of a density of hundreds of millions of tons per cubic inch. Many such neutron stars, whose existence was predicted by Lev Landau, are now known in the form of pulsars; neutron stars that spin rapidly, up to a thousand revolutions per second, and that, as they spin, flash out radio beams, generated by their rapidly rotating magnetic fields, like a cosmic lighthouse.

These neutron stars balance the squeeze of gravity against the spring of the neutrons that object to further compression in just the way that the electrons objected when it was they who had the job of holding the star up against gravity. But if the neutron star is more massive than about two solar masses—the exact figure is not known—then the spring of the neutrons is not strong enough; gravity triumphs and there is now nothing to resist collapse of the entire star into a black hole.

Supernovae. Stars of mass greater than the Chandrasekhar limit do not necessarily slide gently into a white dwarf or neutron star terminal state or even simply collapse into a black hole. To see what might happen I will rewrite the scenarios above in more dramatic form.

Stars are somewhat like onions with different compositions and densities and temperatures in their different layers. Nuclear reactions take place throughout the star and convert the hydrogen, of which the star is originally largely composed, into heavier elements at different rates in the different layers, the more rapidly the deeper the layer since the temperature increases toward the middle. The inner part of a heavy star, its core, may eventually exhaust its nuclear fuel in the sense of converting itself entirely into those elements that are stablest from the nuclear point of view, namely, iron and its immediate neighbors, which, therefore, do not generate energy by nuclear reactions. But the core of the star, following this complete nuclear combustion, is at an exceedingly high temperature and is a source of electromagnetic radiant energy that can photodisintegrate the iron and other nuclei and thereby induce a phase change of the core into a nuclear fluid of alpha particles (^{4}He nuclei) and neutrons. This abstraction of energy from the radiation field drastically upsets the balance of pressures inside the core, which collapses gravitationally, releasing more gravitational energy, which raises the temperature yet further. The temperature is now high enough for the alpha particles themselves to be photodisintegrated, which brings about another phase change of the core into a neutron/proton/(electron) gas, abstracts yet more radiant energy, and induces further gravitational collapse.

The processes I sketched previously now come into effect: the electrons combine with the protons, liberating gigantic numbers of neutrinos and yielding neutrons. All this takes only a few seconds from the initial collapse of the iron-rich core. The outer layers of the onion, still rich in combustible nuclear fuel, fall in upon the collapsed core and themselves ignite explosively at the suddenly and greatly increased temperature encountered in the outwardly moving shock wave resulting from the termination of the collapse of the core. The balance between the energy released in the gravitational collapse of the core, with its emission of neutrinos and its proton-to-neutron conversion, and that released in the subsequent ignition of the in-falling outer layers varies with the type and mass of the star; various forms of supernova are recognized. But the upshot of the explosion is always the blowing-off into space of what were the outer layers of the onion, taking with them the vast range of chemical elements in those outer layers, together with those other elements that have been synthesized in those dramatic last seconds of Nature's ultimate nuclear weapon. The release of energy in these so-called supernovae explosions is stupendous and for a few weeks the debris of the explosion may outshine all the other stars of the galaxy put together. These supernovae are not of very frequent occurrence: in our own galaxy containing about 10^{11} stars there is probably one supernova every thirty years or so. We should be grateful for the supernovae because they are the source of the heavy elements that, having been blown off into space, are eventually collected by other stars and gathered into planets and so turned into you and me: we are all stardust.

The collapse of the inner part of the supernova, together with the effect of the explosion of the outer layers, is likely to produce a neutron star as the residue of the event. Indeed several cases of this type are known where one sees a pulsar at the center of a still-visible expanding mass of blown-off material, as in the Crab nebula, which is a supernova that exploded, as seen from Earth, in A.D. 1054. (Incidentally, the Chinese court astronomers who observed it got into serious trouble for having failed to predict it, which is a bit hard on them.) But it is also not at all improbable that the supernova explosion will compress the inner part of the star, not into a neutron star, but into a black hole, which we should not then, by definition, directly see.

Black Hole Signals. I have just, cautiously, referred to our not seeing black holes directly. By the nature of the beast an isolated black hole would send us no signal of its presence except, in principle, through the wholly imperceptible dribble of its evaporation and, of course, by its gravitational pull. If, however, the black hole is not isolated but is in the neighborhood of material that can be gravitationally sucked into it, then that material may itself radiate radio waves or visible light or X rays or gamma rays in the violent processes accompanying its sucking-in; in this way we may get news of the presence of the black hole, even though we cannot directly receive messages from within its event horizon. It is likely that the energy radiated in this way is a sizable fraction of the entire mass energy of the material falling into the black hole as measured by Einstein's $E=mc^2$; the black hole is a tremendously efficient energy source. Indeed, it is thought that this is probably the explanation of the mysterious quasars, quasi-stellar objects, that are distributed throughout the visible Universe and that shine typically one hundred times more brightly than the brightest galaxies, although their volume is 10^{12} times smaller. A gigantic black hole, of mass perhaps comparable with that of an ordinary galaxy, and situated within one, might be expected to produce just this phenomenon as it continues to gobble up the remainder of its galactic environment.

We can obviously learn about black holes by inference, but some of the inferences are becoming strong. Thus we know of several so-called X-ray binaries, which consist of a visible star circling an invisible companion object, the combined system being a powerful emitter of X rays whose intensity may fluctuate rapidly —on the scale of less than a millisecond. If the measured mass of the invisible object is greater than the maximum possible for a neutron star, namely, two or three solar masses, then the only known candidate for it is a black hole; the rapid fluctuation of X-ray intensity is consistent with the small effective size of the black hole that is sucking in matter from its visible companion and so producing the X rays. Several such binary systems containing black hole candidates are known; the most famous is called Cygnus X-1 where the invisible object is of about ten solar masses.

It is thought that many, if not most, galaxies may have massive black holes, of millions of solar masses, at their centers; there is, indeed, some evidence that our own galaxy may be so endowed.

Naked Singularities? I have mentioned that, as we understand it, at the center of a black hole there must be a singularity at which space and time come to an end and at which the laws of Nature, as we know them, break down. What, then, happens there? We obviously will never know because news cannot come out of a black hole and if we go in to see for ourselves we shall not live to tell the tale. However, although all black holes must have singularities at their center, not all singularities need necessarily be inside black holes: there is the in-principle possibility of the existence of a naked singularity, that is to say, one that is not shrouded from view by an event horizon. If such exist, and we have no idea at the moment how they might be formed in practice in astronomical events now in progress, then we might directly observe the end of space and time and their unguessable consequences. This naked singularity is thought by some to be a rather indecent possibility, and Roger Penrose has proposed a kind of cosmic censorship hypothesis, which says that all singularities must be respectably shrouded from view by an event horizon.

Wormholes and Time Travel. At a singularity, naked or otherwise, the curvature of space is infinite and the equations of general relativity itself have no solution. However, under conditions of large, but not infinite, curvature, general relativity offers bizarre solutions known in the trade as "wormholes" that amount to a kind of short-circuit in space and time linking what are otherwise remote parts of the Universe. It is a moot question whether or not these wormholes are traversible in the sense that one could, in principle, make use of them actually to get rapidly from one part of the Universe to another and/or travel in time and violate causality. The answer depends on a better understanding than we now have of the interplay between general relativity and quantum effects and on the validity or otherwise of the cosmic censorship principle. Within classical general relativity it does not seem that wormholes have much cosmological significance; they are probably, in any case, confined to the interior of black holes, but with quantum gravity, the conversion of general relativity into a quantum theory, it may well be different, and I shall touch upon this later. However, general relativity at least invites us to consider such matters as time travel that are normally regarded solely as the province of science fiction.

Time travel is always fraught with problems such as what happens if you go back in time and murder your own remote ancestors; it may well be that we simply prefer not to think about it. I should perhaps remark that general relativity is not restricted to wormholes in its offerings of time travel: another mechanism was discovered by Kurt Gödel; we must obviously keep an eye on it.

The End of the Universe? Before leaving, temporarily, the fascinating field of black holes and singularities I feel that I ought to point out that there are some who see in them our ultimate danger. They consider that since time and space terminate at a singularity, then if that singularity is naked, and so in communication with the entire Universe, time and space must vanish everywhere and the entire Universe be destroyed. They further argue that although we cannot see how to construct a naked singularity directly, the Hawking evaporation of a black hole must eventually lead to one as the event horizon shrinks to zero so that the evaporation of a black hole would destroy the Universe. It would obviously be up to us to keep shoveling matter back into the black hole to keep it from completely evaporating; that is, of course, if we thought it was worth doing.

While in apocalyptic mood I might put together two of the recent points that I have been making and look at our ultimate fate if we escape all the other possibilities, some of which I have already mentioned and some of which I shall mention shortly. I have said that iron is the most stable form of ordinary matter. So eventually everything that has escaped other fates will turn to iron. According to Freeman Dyson the necessary quantum tunneling processes will take about 10^{1500} years. And then everything will collapse into black holes after about 10^x years where $x=10^{26}$. Beyond that I do not think that we need bother very much.

3. The Big Bang

Having taken a preliminary look at how it might all end, I will now take a look at how it all began. We think that there was, in fact, a beginning to our Universe, about 15,000,000,000 years ago, and very spectacularly at that, in the Big Bang. There are three chief reasons for our thinking that there was a Big Bang before which there was nothing, neither space nor time nor matter, and from which everything that we call our Universe evolved.

The Recession of the Galaxies

The first reason is that when we look out into space we see it as populated with galaxies, some very like our own and some of different construction, but all rushing apart with a velocity that is proportional to their distances from each other. This is exactly what we should expect if there had, indeed, been a Big Bang at the beginning of it all that had flung out matter and space together—not flung out matter *into* space because there was no space before the Big Bang into which matter might be flung; remember the intimate relationship between matter and space that general relativity specifies. This remarkable expansion of the Universe was announced by Edwin Powell Hubble in 1929 and is known as the Hubble recession.

The Microwave Background Radiation

The second reason for believing in the Big Bang is that the entire Universe appears to be permeated with electromagnetic radiation of an extraordinarily uniform distribution—isotropic to better than 0.01%—and corresponding quite accurately in its spectrum to what one would expect for a body glowing at a temperature of about 3°K—3 degrees above the absolute zero. This is called the microwave background radiation; it was discovered by Arno Penzias and Robert Wilson in 1965. As had originally been pointed out by George Gamow in 1948, this is just what is to be expected from a Big Bang at the beginning of time that produced a large amount of radiant energy that subsequently cooled as the Universe expanded. (It is amusing to note, in passing, that since no part of the natural Universe can ever have been at a lower temperature than its present 3°K, Nature can never have carried out those lower temperature experiments that are commonplace in our terrestrial laboratories.)

The Cosmic Abundance of the Light Elements

The third reason for believing in the Big Bang is the cosmic abundance of certain of the light elements, particularly ^{4}He but also ^{2}H, ^{3}He, and ^{7}Li. By cosmic abundance is meant the universal distribution in the cosmos of elements whose presence there we cannot understand as being due to the processes subsequent to the formation of galaxies and stars and that we therefore believe must have been there from the beginning.

^{4}He is found to be cosmically abundant to the tune of about 25% by weight, the great bulk of the rest being ordinary ^{1}H. ^{4}He is a tightly bound nucleus that is difficult to destroy in the stellar furnaces and it is also not possible to think up ways of synthesizing it there to any degree that would be significant on the cosmic scale. We must conclude, therefore, that this 25% abundance represents its manufacture in the Big Bang itself. It is quite easy to work out how much to expect.

When the first instants of the Big Bang are passed, as I shall discuss shortly, we must presume that we shall be left with neutrons and with protons in rather accurately equal abundance since they are almost identical particles apart from their electrical charge and electrical forces will have been quite unimportant at the high energies corresponding to the high temperatures earlier obtaining. But neutrons are, in fact, very slightly heavier, by about 0.1%, than protons, and so the nuclear reactions that interconvert them, mediated by electrons and neutrinos, push the balance toward protons and deplete neutrons. If these reactions are sufficiently vigorous to keep the situation in thermodynamic balance, the preponderance of protons over neutrons, as a function of temperature, is given by a simple chemical expression, the Saha equation.

This thermodynamic balance is maintained until the Universe has cooled to such a temperature and the density has become so low that the reaction rates of neutron-proton interconversion become negligible. We are then left with a preponderance of protons over neutrons to a degree that depends on the temperature at which this decoupling takes place. This temperature itself depends upon the reaction rates as a function of temperature, which are well known from terrestrial experiments, and the rate at which the temperature and the density have fallen, which is specified in an essentially model-free and parameter-free way by the general theory of relativity. This decoupling we calculate to take place about a second after the Big Bang. We can therefore rather confidently predict the proton/neutron excess at this stage of the proceedings.

The next stage, as the Universe continues to expand and its temperature continues to fall according to the known law, takes the form of various nuclear reactions between the neutrons and the protons and those heavier elements to which those reactions give rise. We can also measure these various reaction rates in the laboratory or infer them from such measurements. We can calculate, therefore, what happens to the brew as time goes by and the lighter elements are gradually cooked into the heavier ones as the temperature of the oven falls. When the temperature has fallen to the point where no further nuclear reactions take place, because the energy of thermal agitation of the nuclei is not great enough to overcome their mutual electrostatic repulsion, the final elemental mix is attained.

For some time after the one-second mark, at which this stage of nucleosynthesis sets in with the neutron/proton mix that I have just discussed, nothing much happens in the way of building up a significant amount of the heavier elements because they get broken down, photodisintegrated, by the electromagnetic radiation that accompanies the high temperature. But after about a minute, formation of the elements beats their photodisintegration. After three or four minutes it is pretty well all over; the temperature is too low to sustain the nuclear reactions at any significant rate and we are left with the mix of elements out of which the Universe must then construct itself; the rest is history.

There is, in fact, one additional parameter that I have not yet mentioned and that is important in the calculation of the abundances of some of the elements produced in these first few minutes. This can be expressed in terms of today's mean density of nucleons in the Universe as a whole (including those that have disappeared down black holes). We do not know this mean density very well, as I shall discuss shortly, but the important matter of the theoretical cosmic ^{4}He-to-proton ratio, about 25% by weight experimentally, as I mentioned earlier, is very insensitive to this mean nucleon density and, indeed, comes out to be just about the observed 25%. The other abundances do depend more sensitively on the mean nucleon density, some of them rather violently, but all agree with observation for a value for that density that is very reasonable and does not conflict with any known data. To this I shall return. Our understanding of matters of nucleosynthesis in the Big Bang, and subsequently in the stars, is the result of contributions from many persons since the mid-1950s, prominent among them being William Fowler and Fred Hoyle.

Degrees of Freedom; Neutrinos

Before leaving the synthesis of the light elements in the Big Bang, note that the rate of cooling of the early Universe, upon which all this depends, itself depends on the so-called number of degrees of freedom among which the energy of the system is shared out: the more degrees of freedom the more rapid the cooling.

The number of degrees of freedom is, crudely speaking, just the number of distinguishable ways in which the particles present can be identified, and this depends on the number of distinguishable species of particle and the number of ways in which each species of particle can distinguish itself: for example, the photon has two degrees of freedom because it can be right-hand polarized or left-hand polarized. In particular, the effective number of degrees of freedom depends on how many different species of neutrino there are whose masses are sufficiently small in relation to the thermal energy corresponding to the temperature at the time in question, namely, of the order of 1 MeV.

We know of three species of neutrino, namely, those associated with the ordinary electron and with the two heavier species of electron, namely, the muon and the tauon. We know that the masses of the electron and muon neutrinos are sufficiently small and suspect that that of the tauon neutrino is also. Indeed the excellent agreement between Big Bang theory and the observed cosmic abundance of ^{4}He is found for two or, better, three species of neutrino while four could be squeezed in only very uncomfortably. We have no idea why there are three different species of electron with their separate neutrinos, but the Big Bang seems to be telling us that it is very unlikely that there is more than one other species, at least with a light neutrino, and that there are most probably only three in all.

The Three Evidences

These three evidences: the expansion of the Universe as signaled by the mutual recession of those marker particles within it, the galaxies; the 3-degree microwave background radiation; the agreement with expectation of the cosmic abundances of the lightest elements, point compellingly toward the Big Bang origin of the Universe.

The Cosmological Constant

It is of interest, in passing, to make two important observations at this point. The first is that Einstein's theory of general relativity automatically provides for an expanding Universe. You do not have to put in the expansion by hand or by some special mechanism. In fact, the trick is to stop the Universe from expanding if you do not want it to. Einstein originally did not want his Universe to expand: all the thinking of the age was in terms of an essentially static Universe, so he introduced into his equations a legitimate but "unnecessary" term called the cosmological constant whose effect could be to balance out the otherwise intrinsic tendency of the Universe to expand. Einstein came to regret this, but in truth, such a term cannot be excluded from the equations if they are to be complete in the range of possibilities that they offer within the overall general relativistic philosophy. I shall return to the cosmological constant, which has become one of the greatest mysteries and most hotly debated issues of modern cosmology.

The Big Bang as Singularity

The second observation in passing concerns the extreme nature of the Big Bang. I have already mentioned the Penrose theorem that every black hole has at its center a singularity of zero size and infinite density at which space and time stop and beyond which it makes no sense to attempt to penetrate in terms of the laws of physics. Now black holes are formed by the infall of matter. What we see in the expanding Universe is the opposite of this: the apparent ejection of matter. In 1970 Stephen Hawking asked what happens if you take the expanding Universe and run time backward to trace its earlier evolution according to general relativity.

Hawking and Penrose discovered that things must have started from just such a singularity as is at the center of a black hole. In classical general relativity there is no way of starting off an expanding Universe other than from a singularity: you cannot have a Big Bang starting with an object of finite size.

In a sense this is comforting because just as space and time end at the singularity inside the black hole, so space and time start at the singularity of the Big Bang and this avoids having to answer the embarrassing question of what happened before the Big Bang. St. Augustine put it very clearly:

What did God do before he made Heaven and Earth? I do not answer as one did merrily: He was preparing Hell for those that ask such questions. For at no time had God not made anything because time itself was made by God.

This is, indeed, quite close to the position of general relativity.

The Planck Time

However, and I stress this, what I have just said is in the context of classical general relativity, which, as I have mentioned before, is not a quantum theory but which must at least be brought into some sort of cohabitation, preferably a respectable one, with the quantum theories of the other forces that, together with gravity, determine the course of Nature. In particular, we must believe that even under a gravity-dominated regime, that most fundamental of quantum regulations, Heisenberg's uncertainty principle, will apply.

To illustrate the collision between general relativity and the quantum theory, consider the instants following the Big Bang singularity. At a given time, t, after the Big Bang, that is to say, after the beginning of time, the observable Universe, which is all that we can, by definition, meaningfully discuss, will consist of a sphere of radius, R, equal to the time in question multiplied by the velocity of light, c, so $R = ct$. But, as I mentioned when discussing the synthesis of the light elements following the Big Bang, general relativity prescribes the relationship between the various variables of interest such as the time, the temperature, and the density, ρ, whatever form that density might take. The equation of immediate interest is $32\pi G\rho t^2 = 3$, in which G is the constant, Newton's gravitational constant, to which the gravitational force is proportional. Now if the density is p then by Einstein's famous $E = mc^2$ relationship between energy and mass, the energy density must be ρc^2 and the total energy content of the Universe is $E = 4\pi R^3 \rho c^2/3 = c^5 t/(8G)$.

The appropriate uncertainty relationship, named after Werner Heisenberg, that the quantum regulation requires is what specifies the necessary imprecision in our knowledge of the energy content of a system that is entrained by the fact that our measurement of that energy content has taken only a certain length of time. Thus quantum theory does not permit us to state that, as in the equation above, the energy content of the Universe at time t is precisely $c^5 t/(8G)$ because we have had only time t in which to measure it and our knowledge must therefore be imprecise by an amount $\hbar/t$, where $\hbar$ is Planck's constant, which characterizes all quantum statements. There will, therefore, be a time, t_P, called the Planck Time, after the Big Bang at which our knowledge of the energy content E of the Universe must be unsure by an amount equal to E itself. This is like having an idea that your bank balance is about $1000 but having to recognize that it might in fact be $2000 or so, or, equally probably, that you might be in overdraft; such information is obviously not of very great use. The Planck Time, t_P, at which the equality between the energy content and our uncertainty of it is struck, is then given by $c^5 t_P/(8G) = \hbar/t_P$ or $t_P = (8G\hbar/c^5)^{1/2}$ which, putting in the values of the constants, is about 10^{-43} seconds.

At later times the uncertainty in the energy content entrained by the quantum nature of the world becomes less and less in relation to the energy itself and we can feel more and more comfortable in using general relativity without worrying about the quantum restrictions. But at times earlier than the Planck Time the situation gets worse and worse and we literally do not know what we are talking about.

The Compton Wavelength

This irreconcilable collision between classical general relativity and the quantum theory can be illustrated in a complementary way. At the Planck Time the observable Universe has a radius $R_P = ct_P$, namely, about 10^{-33} cm (which is that smallest distance that comes into our present picture of the Universe that I mentioned when discussing our own size in relation to sizes that we were led to consider within the physical Universe). R_P is called the Planck Length and is the smallest distance that we can meaningfully discuss, just as t_P is the earliest time that we can meaningfully discuss. At time t_P the Universe also has a mass called the Planck Mass $m_P = \hbar/(t_P c^2)$, which is about 10 micrograms or 10^{19} GeV in energy terms.

Now a particle of mass m has an energy equivalence mc^2. If we consider only a short interval of time Δt, then the Heisenberg uncertainty principle that we have just used in connection with the earliest instants of the Universe says that I cannot determine the energy content of the system that I am examining to better than $\Delta E = \hbar/\Delta t$, so in that sense I can consider the energy ΔE as available for me to "borrow" for the time Δt because there is no possible way of checking, within the time available to make that check, namely, Δt, whether I have got it or not. So out of this short-term borrowed energy ΔE I can make a particle of mass m and so energy mc^2, namely, $mc^2 = \hbar/\Delta t$, and in that time Δt the particle can travel a distance of order $c\Delta t = \hbar/(mc)$, which is therefore a kind a measure of the uncertainty of its position. This length $\hbar/(mc)$ is called the Compton wavelength of a particle and enters intimately into all quantum discussions of particle properties and behavior. (If you found the illustration of the uncertainty in your bank balance useful in thinking about the Planck Time you might also find it useful to see in the borrowing of energy for a limited period of time that the quantum theory allows—the more energy you borrow the quicker you have to pay it back, $\Delta E = \hbar/\Delta t$—a parallel with embezzlement where you can get away with embezzling small sums for a long time, but large amounts tend to get you found out more quickly.)

So in the case of the earliest instants of the Universe there will be a quantum uncertainty in its location given by its Compton wavelength at that time, namely, $\hbar/(mc)$, where the mass is, following the earlier discussion, $m = c^3 t/(8G)$ so that the Compton wavelength is $8\hbar G/(c^4 t)$. There will come a time when the uncertainty in location of the Universe is equal to its actual size, $R = ct$. Equating R and the Compton wavelength shows this time to be just the Planck time $t_P = (8G\hbar/c^5)^{1/2}$ that we earlier derived. At earlier times the uncertainty in position of the Universe is greater than its own size and at later times things settle down to a classical picture. A verbal statement of this is that the Planck Time is the time in the development of the Universe at which the curvature of space is equal to the Compton wavelength.

Quantum Gravity

All this has been to emphasize the irreconcilability of classical general relativity and the quantum theory when the two come into conflict. We simply have no mechanism for discussing what happened at times earlier than the Planck Time and shall not have such a mechanism until general relativity has itself been put on a quantum basis. The search for such a theory of quantum gravity is brisk; several candidate theories have been put forward but none has yet proved to be unexceptionable. More about this later.

In the meantime, there is a faint glimmering of a connection through the one application of quantum theory that has been made in a gravitational context, namely, the evaporation of black holes by the Hawking process that I mentioned earlier. I then stated that the lifetime of a black hole of mass 10 micrograms against such evaporation would be about 10^{-43} seconds. Now 10 micrograms is, as we have just seen, the Planck Mass and 10^{-43} seconds is the Planck Time, so a black hole of the Planck Mass would evaporate at the Planck Time, which emphasizes the connection between the singularity at the center of a black hole and the time-reversed singularity at the beginning of the Universe. Appropriately, black holes at times before the Planck Time, and therefore of mass less than the Planck Mass, would evaporate in less than the Planck Time, and this has led Hawking to picture the time earlier than the Planck Time as a kind of seething foam of forming and evaporating black holes. I hope you find that useful.

Primordial Black Holes

Black holes of small mass, less than a few solar masses, cannot form by normal mechanisms of stellar evolution and gravitational collapse. But they can form under high compressive forces such as must have existed shortly after the Big Bang on account of the lack of perfect uniformity in the density distribution. We know that such variations of density must have existed, to some degree, in the very early Universe to account for the fact that the visible matter of our present-day Universe is largely clumped into galaxies. There is, therefore, a strong likelihood that shortly after the Big Bang such primordial black holes must have formed to some degree. Those with masses of less than about a thousand million tons would have evaporated already by the Hawking mechanism that I have already discussed, but those of about this mass would be in their final death throes today; we should see them through the X rays and gamma rays that they would be emitting. No such radiation has been definitely detected although it is being sought for energetically.

There is, in fact, a substantial general uniform gamma-ray background that contributes to the flux of cosmic radiation coming in from outer space, but it does not have the spectrum to be expected for black hole evaporation. It does, however, enable one to make statements limiting the overall density of primordial black holes left in today's Universe: it is not very great. Of course, primordial black holes would probably tend to be attracted into galaxies and our best chance of observing them is probably not through the summed radiation from a very large number throughout the Universe, such as would contribute to the uniform gamma-ray background that I have just mentioned, but from the final near-explosion of individuals within our own galaxy as they come to the end of their lives.

You might, incidentally, draw some comfort from this discussion of the evaporation of primordial black holes in the context of my earlier apocalyptic suggestion that the evaporation of a single black hole would lead to a naked singularity that might destroy the Universe. If the primordial black holes have been evaporating ever since the Big Bang we are probably safe.

Two Important Problems

The Big Bang seems to work very well as a kickoff to the Universe. But the story as I have told it so far immediately encounters two important problems.

The Photon-Nucleon Ratio. The first problem concerns the overall brightness of the Universe, that is to say, the amount of electromagnetic radiation that it contains in relation to the amount of matter as measured by the number of nucleons. The electromagnetic radiation of the Universe is found overwhelmingly in the microwave background radiation that I mentioned as one of the three chief evidences for the Big Bang: there are about 10^9 photons of this radiation for each nucleon. This is an almost incredibly large number and can be compared with the brightness of that brightest of the well-understood cosmic phenomena of our present-day Universe, a supernova, which emits only about one hundredth as many photons for each of its nucleons. The Universe as a whole is of a staggering brilliance. As I explained earlier, this dazzling microwave background radiation is the fossil remnant of the Big Bang itself; those 10^9 photons per nucleon have been around ever since and so represent the situation as soon after the Big Bang as things had settled down into the constituents of our present-day Universe. What should we, offhand, expect by way of the photon-to-nucleon ratio resulting from the Big Bang? There is nothing in our knowledge of nuclear and particle physics to lead us to expect that any one sort of particle would have been especially favored over any other; there is a kind of cosmic democracy among particles, particularly at the very high energies then of concern. We might have expected some slight variations between the particle species, by a factor of ten or possibly by as much as of a hundred, but a factor of 10^9 is out of the question and the brightness of the Universe demands a fundamental explanation.

No Antiparticles. The second problem encountered by the simple Big Bang story is that the Universe, so far as we can tell, is made exclusively of matter, ordinary nucleons and electrons, with no antimatter in the form of antinucleons and positrons. (Of course which is matter and which is antimatter is solely a matter of convention.) Small numbers of antiparticles are found in the cosmic radiation, but these can be satisfactorily accounted for as having been made in the collisions between the ordinary particles of the cosmic radiation and other ordinary particles upon which they have impinged in the course of their passage through intergalactic or intragalactic space or through the Earth's atmosphere. (The cosmic rays themselves range in energy up to 10^{20} eV or so and are chiefly protons, together with a few heavier nuclei, that have been raised to these gigantic energies—exceeding by a factor of 10^8 those reached by the most powerful Earthbound accelerators operating today—by various forms of celestial electromagnetic machines, most probably pulsars and quasars.)

The expectations of cosmic democracy are much stronger in relation to particle versus antiparticle production than they were in relation to the approximate equality of production of the various species of particle that we have examined as our first puzzle left over by the Big Bang. In fact, physics-as-we-understand-it is rigorous in its requirement that the initial situation, following any form of creation that we know how to incorporate into our most general philosophy, should be one of exact equality between numbers of particles and their antiparticles without any differences such as might be thought to be due to statistical fluctuations and so on. What then might be the source of the observed total lopsidedness of today's Universe? Was it, in fact, simply made that way? If that were the case we might as well give up trying to understand anything to do with the very early Universe and regard its initial composition as a matter of fiat. Or have all the antiparticles gone somewhere and, if so, where?

Cosmology and Particle Physics

At this point particle physics shakes hands with cosmology and enters into a partnership that is proving to be of intense interest and value to both. I will tell the story as according to one simple and explicit model although it can be told with greater generality to the same end. The story does not begin at the beginning, namely, the Big Bang itself, because, as I have explained, we do not know how to begin to talk in quantum mechanical language until after the Planck Time of 10^{-43} seconds when gravity begins to slacken its grip and permits us to get on with physics-as-we-understand-it-better.

The Four Forces

What are the forces that, after the Planck Time, then begin to shape the nascent Universe? As I mentioned earlier, in our present everyday Universe we are aware of, apart from gravity, three quantum forces, all very different-looking from one another, namely, the electromagnetic force that is described through quantum electrodynamics (QED), the strong force within the nucleon and the nucleus that is described by quantum chromodynamics (QCD), and the weak force of beta radioactivity.

These forces have very different apparent strengths: if we hold two protons at a distance apart equal to their own size, namely, about 10^{-13} cm, the largest force between them is the strong or nuclear force—it amounts to about 3 tons' weight; the electrical force is then only about 60 pounds, while the weak force is a mere one hundredth of an ounce. (The gravitational force between the protons is unimaginably smaller still—some 10^{38} times less than the strong force. Gravity is really very weak—a hydrogen atom held together by gravity would be bigger than the observable Universe, and yet, because of the fact that it acts indifferently among all objects irrespective of their electrical charge and other properties, and because of its long range, it dominates the large-scale structure of the Universe and ultimately wins out over the other forces, controlling both the end of things in the black holes and the beginning of things in the Big Bang.)

The three quantum forces differ also in the ways in which their effects fall off with distance. The familiar electrical force falls off with the inverse square of the distance between the interacting charges (as also gravity does), and this is due to the fact that its messenger particle, the photon, whose exchange mediates the force, has zero mass. The weak force is mediated by the exchange of the W and Z particles that I have already mentioned. They are heavy, about one hundred times the mass of a proton, and so, by the mechanism of embezzlement that I used when introducing the notion of the Compton wavelength, these particles can carry the weak message over only very short distances before they have to deliver it by being absorbed by the other particle, namely, about 2×10^{-16} cm, so the force appears to be weak, not so much because the message is itself not clear and strong, as because it does not carry very far.

The strong force between the quarks out of which neutrons and protons are constructed is also mediated by a massless particle, the gluon, and so also falls off with the inverse square of the separation, like the electric force. But there is a dramatic difference between the strong and the electromagnetic forces, which is that whereas the electric messenger, the photon, does not interact with itself through the force that it propagates, the strong messenger, the gluon, does so interact. This fact that there is a gluon-gluon interaction, as well as a quark-gluon interaction, leads to a progressive strengthening of the effective quark-quark force as the separation between the quarks increases, much as the tension in a rubber band increases as you stretch it, and this strengthening is superposed upon the inverse square weakening as the separation increases and, at large distances of quark-quark separation, totally dominates it. This is clearly related to the phenomenon of confinement: the fact that although the nucleon contains three quarks, free quarks apparently cannot be liberated from it. Confinement is not yet fully understood, and we do not yet know if it is absolute, but the increase of the quark-quark force with distance, which is understood, points in the right direction.

(It must not be thought that this catalog of four forces is necessarily complete; there may be other forces, perhaps many others, of which we are not yet aware either because they are very weak or because they are strong but act only between particles, themselves not yet discovered, that themselves interact only weakly with those with which we are familiar. In recent years there have, indeed, been persistent reports of a fifth force that acts something like a gravity of short range whose effects drop off rapidly after a few hundred meters; it now seems to be going away, but there is no reason why such forces should not exist; perhaps without knowledge of them we shall not be able to construct a fully self-consistent story of the interrelationships between the forces of which we do have knowledge in the sense that I shall now describe.)

Unification of the Quantum Forces

The three quantum forces that, together with gravity, shape, so far as we know, our natural world, look very different from each other. But there are similarities. In particular, the messenger particles, the photon, the gluons, and the W and Z particles, are all of intrinsic spin unity in the appropriate quantum units; the theories of the interactions are all so-called gauge theories. So it is natural to ask if there might not be connections between the forces even though they look so very different. Such quest for unification of our understanding makes a very strong appeal and has always done so since the days when Newton unified mechanics through bringing together gravitational and inertial masses and since Maxwell, following Faraday, unified electromagnetism through bringing electrostatic and magnetic phenomena under a common description. Einstein spent his last decades in a fruitless search for the unification of gravity and electromagnetism, a search that we now recognize cannot succeed until we understand the quantum aspects of gravity.

We have, however, now achieved a large measure of unification of the electromagnetic and the weak forces through the work of Sheldon Glashow, Abdus Salam, and Steven Weinberg. We understand how these forces can be placed into a unique relationship with one another that falls short of complete unification only because it contains a parameter, the so-called Weinberg Angle, that measures the relative intrinsic strengths of the two interactions and that is not itself specified by the theory. The very different apparent strengths of the electromagnetic and weak interactions is indeed seen, as I foreshadowed earlier, as due to the very different ranges of the forces, not to their difference in intrinsic strength, which difference is now seen to be very little.

Electroweak Unification

We now speak, therefore, of the electromagnetic and weak forces, not as distinguished interactions, but as different aspects of a single interaction, the electroweak interaction, that operates differently in different contexts. Thus in the context of beta decay, which changes a neutron into a proton, only the weak aspect of the electroweak interaction will operate because a change of charge is involved that cannot be effected by the electrically neutral photon but that can be effected by the electrically charged W particle. However, in the interaction between, say, an electron and a proton within an atom, both the electric force, carried by the photon, and the weak force, carried by the electrically neutral Z particle, can contribute. But since the effect of the weak force, although that force is intrinsically of strength comparable to the electric force, is, in practice, several million times less than that of the electric force because its range is so much shorter, the exchange of the Z messengers within the atom has only a minute effect on the actual atomic structure as evidenced through, for example, the energy levels. This minute effect can, however, be detected because of another significant intrinsic difference between the forces.

Parity

This other intrinsic difference between the weak and electromagnetic forces is that whereas the interaction between photons and electrical charges is completely indifferent in regard to whether the photons are right-handed or left-handed circularly polarized, the weak interaction between the W or Z particles and other particles is intrinsically of only a left-handed nature; that is to say, the transactions between a W or Z particle and another particle or system is like that of a left-handed corkscrew so that this handedness is transmitted to the products of the interaction. The interaction simply does not work with the opposite handedness any more than you can open a bottle of wine with an ordinary right-handed corkscrew by turning it anticlockwise. We do not understand why this asymmetry should exist and why there are not also right-handed W and Z particles. Perhaps there are, but we infer experimentally with good confidence that if the right-handed W particle exists, then it is at least five times as heavy as the left-handed W particle so that its reach is more than five times less and its influence correspondingly more feeble.

In the jargon, the ambidextrous electromagnetic interaction conserves parity and the left-handed weak interaction does not. So when we look at the structure of an atom, which is held together predominantly by the parity-conserving electromagnetic interaction, but with a tiny contribution from the parity-nonconserving weak interaction, we cannot hope to see the effect of the latter on the energy level structure itself because that does not reflect any handedness: the levels are simply at certain energies and these are only infinitesimally shifted by the weak interaction from the positions they would have occupied without it. But if we look at the radiation, the light, emitted when the atom makes transitions between its energy levels, then that tiny bit of handedness can show up as a very tiny degree of circular polarization of the emitted light, that degree being of the order of a part in a million or so.

That is a very small degree, but it can be, and has been, measured in a few cases and its magnitude checks out well with what we expect from the electroweak unification.

Analogous experiments have revealed the electroweak mixture at work in the disintegration of the deuteron by longitudinally polarized electrons where the predominantly electromagnetic disintegration has superimposed upon it a tiny bit of the weak interaction so that the probability of such disintegration depends very slightly on whether the electron's spin points in the same direction as its motion, and the electron therefore behaves like a right-handed corkscrew from the point of view of the weak interaction, or whether the spin points in the opposite direction, in which case the electron is like a left-handed corkscrew. Again, the very slight experimental dependence upon the direction of the electron's polarization checks out well with the electroweak theory.

The head-on collision between energetic electrons and positrons can give rise to annihilation with a pair of oppositely charged muons or tauons as products: if this were a purely electromagnetic process, we should expect the product particles of a particular sign of their electric charge to be emitted equally forward and backward, but the weak contribution to the process generates a forward-backward asymmetry in the distribution of just the expected amount.

There is no such thing as a purely electromagnetic process: there is always an accompanying weak contribution whose magnitude stands in a fixed relationship to the electromagnetic as prescribed by the electroweak unification.

Spontaneous Symmetry Breaking

All this is very strange. How can the electromagnetic and weak interactions be unified, be different aspects of the same thing, when their messenger particles, the photon for the electromagnetic interaction, and the W and Z particles for the weak interaction, are so very different: the former is of zero mass and the others are one hundred times more massive than the proton? Indeed, in our everyday world the two interactions are very different just because the energies usually encountered are much less than correspond to the masses of the W and Z particles so that the mass difference, as reflected through the range of the forces, is tremendously important: the photon can reach out much farther than the W and Z particles. If we went to very high energies, significantly higher than correspond to the masses of the W and Z particles, say, to a thousand times the mass of the proton, then the difference would disappear because the high energy would mean that we would not have to rely primarily upon embezzlement to bring the W and Z particles into being and so their reach would be effectively extended.

This would be the situation if the world were at a very high temperature such as it was shortly after the Big Bang—the Universe would, in fact, be at temperatures greater than that corresponding to the masses of the W and Z particles until about 10^{-10} seconds after the Big Bang. So in this very high-energy, very high-temperature situation we would have a complete symmetry between the electromagnetic and weak forces; they would display similar operational strengths and we should have no call to use the term "weak." The two aspects of the single electroweak force would still be distinguished by the handedness transmitted by the weak aspect (unless the temperature were so high that the putative right-handed W and Z particles were effectively equally as abundant as the left-handed), but they would indeed simply be two aspects of the same force. As the temperature falls through the region corresponding to the masses of the W and Z particles, the two aspects largely split apart operationally, although, as we have seen, their effects can still be quantitatively related to one another.

In the jargon, the symmetry present at the higher energies or temperatures is spontaneously broken through the mass differences between the messenger particles. This spontaneous breaking of the symmetry, which so dramatically reveals the enormously different masses of the two sorts of messenger particle belonging to the two aspects of the force, is quite tricky to put onto a respectable basis in the sense of endowing the overall picture with that renormalizability that I mentioned in connection with quantum electrodynamics, which enables us to carry out calculations without running into infinites in our predictions.

The electroweak symmetry breaking can be carried out by recourse to a stratagem due to Peter Higgs (and possibly endorsed by higher authority although that is not yet clear). This involves the existence of a further particle, the Higgs particle, that is associated with the messenger particles of the electroweak interaction; it is a so-called scalar particle having no intrinsic spin or distinguishing features apart from its mass, which is unknown and is not predicted by the theory, and its interactions with other particles, which are also not uniquely specified.

At high energies, well above the W and Z particle masses, the symmetry is not broken and so Higgs is not needed: the wave function or field that represents the Higgs particle has therefore the value zero. But as the symmetry breaks and the two aspects split apart, the Higgs particle must emerge in the sense that its field must assume a finite value. This is the simplest version of the story of spontaneous symmetry breaking, and if it is correct, the Higgs particle must exist. It is obviously quite heavy because it has not shown up yet, but it cannot be too heavy or undesirable consequences are entrained. The search for the Higgs particle will obviously become very hot when accelerators entering new energy regimes become available; it is one of the current Holy Grails of particle physics. As we shall see, similar considerations and particles are critical for our scenarios surrounding the Big Bang.

If the Higgs particle does not exist, then we should have to understand spontaneous symmetry breaking through some sort of synthetic Higgs particle that would do the job of breaking the symmetry but that would not itself necessarily have to be capable of being evidenced as a well-defined particle in the laboratory. Such are the various dynamic symmetry breaking or Technicolor theories that regard the Higgs particle replacement as composites of other particles themselves not yet discovered and probably somewhat embarrassing in some of their implications.

The Higgs Field-Energy Relationship

We must wait and see. But in the meantime, we should remark a most extraordinary property of the Higgs field. One normally expects the energy associated with a field to increase with the field; thus the energy of an electrostatic field increases as the square of the field. However, in the case of the Higgs field, this is not so, and the situation of minimum energy is found for a finite value of the field. As I have remarked, before the symmetry breaks, the value of the Higgs field is zero, but after the breaking it will tend to assume a finite value, then carrying with it a lower energy than for a zero value. There is, therefore, in the Higgs field a kind of explosive tendency to which I shall return.

We might also note here that as we cool the system through the temperature corresponding to spontaneous symmetry breaking, above which the value of the Higgs field was zero, we might well expect, as with many other physical systems with which we are familiar, such as the cooling of water through its freezing point, that the Higgs field might remain at zero for some time past the symmetry-breaking point, even though it were energetically preferable to assume a finite value, just as water often supercools before finally making its phase change to ice. In this case the eventual phase change as the Higgs field assumes its finite value corresponding to its lowest energy could be abrupt and would be accompanied by whatever energy release was implied by the difference between the two energies for zero and finite field values.

The Arbitrariness of Spontaneous Symmetry Breaking

The other most important point to note before leaving spontaneous symmetry breaking is that there is an arbitrariness about it. There is a useful analogy here with a ferromagnet. A ferromagnet is composed of atoms that are held together by the electromagnetic interaction; this acts indifferently in all directions in space; it has no preferred direction. At temperatures above the Curie Point the magnet is indeed not a magnet and shows no preference for any particular direction in space. But below the Curie Point it becomes energetically profitable for the elementary magnets, which are represented by the magnetic moments of the individual atoms and which previously were all higgledy-piggledy, to line up with each other in a particular direction. But that direction is itself arbitrary and is arrived at by accident in a way that cannot be predicted ahead of time. The symmetry above the Curie Point has been spontaneously broken below the Curie Point, resulting in a situation characterized by an infinite and unpredictable variability from magnetization in one direction to magnetization in the completely opposite direction.

So also with the spontaneous breaking of the electroweak symmetry and the emergence of the Higgs field. Technically speaking, the Higgs field, when it becomes nonzero, is characterized by a phase, any value of which would do just as well as any other but which cannot be predicted before the symmetry breaking has taken place. We can consider this phase angle as the analogue of the direction of spontaneous magnetization, below the Curie Point, of the ferromagnet. But just as the ferromagnet behaves differently from the point of view of the external effects that it produces, for example, the magnetic field at a particular point, depending on the way in which the symmetry has in fact chanced to break, so the effect of the breaking of the Higgs symmetry will be different depending on what phase the field, after the breaking, happens to have chosen. To such questions I shall return more seriously shortly.

Buridan's Ass

At this point, I am afraid that I cannot resist retailing the familiar story of Buridan's Ass as a parable of spontaneous symmetry breaking. The essence of spontaneous symmetry breaking is that a system may find itself completely symmetrically located with respect to the coordinate that determines its energy as a function of its position but in which its energy is greater than for some other choice of coordinate, in which that symmetry would be lost. The simplest example is that of a pin balanced upon its point; its situation is symmetrical but obviously unstably so; the slightest displacement will cause it to fall, but one cannot say in which direction: the arbitrariness of the symmetry breaking. But Buridan's Ass is more fun.

Jean Buridan lived in the first half of the fourteenth-century. He was Rector of the University of Paris, a precursor of Descartes and Galileo and a proponent of causality and determinism; scientifically a good thing. He was also reputed to be the lover of not one but two queens of France, Joan of Navarre and Margaret of Burgundy; evidently heads of universities have come down a notch or two since then; however, I understand that modern scholarship has disproved the story although it beats me how. The story of his ass is that it was very hungry and found itself exactly midway between two precisely equally succulent bales of hay and, ass that it was, starved to death because its situation was completely symmetrical, there being nothing that could lead it to prefer one bale of hay over the other. We may be very sure that a real, as opposed to a philosophical, ass would have succumbed to whatever minute fluctuation suggested to it that the one or the other bale of hay should be privileged to break the classical symmetry; but note, was it the one or the other? There is the arbitrariness that we always encounter in spontaneous symmetry breaking. (In point of fact, this story is not to be found in the writings of Jean Buridan; it was invented by his enemies to discredit him. It does, however, occur in connection with Aristotle's "De caelo," which I have already referred to in another context; there it is not an ass but a dog, but apart from that detail the story is correct and is a good illustration of the Latin tag *"liberum arbitrium indifferentiae."*)

Symmetry Breaking in Art

Although it may be thought that a perfect symmetry is more desirable because of that perfection, this is not so. Perfect symmetry is balanced and sterile as are, striking as they may be, the infinite tessellations of Moorish decoration. The unnecessary but deliberate breaking of symmetry is seen in the art of many "primitive" peoples (the quotation marks are deliberate), from the Inuits of the Pacific Northwest through the denizens of the Sepik River of New Guinea to the aborigines of Australia and the Maoris of New Zealand, all of whom systematically turn away from an obvious symmetry and, while preserving an emotional and aesthetic balance in the resultant whole, break the original symmetry with dramatic and thought-provoking effect. Pre-Columbian textiles from Peru often have colorings that are inconsistent with their structural symmetry: they make one wonder. A broken symmetry has a much greater potential and richness than a perfect one; we should be happy that Nature has also taken that route. (Human beings and Nature.)

Perfect symmetries, for the most part, simply and restrictively tell you what you may not do:

> At that cold moralist I hotly hurled
> His perfect, pure, symmetrical, small world.

as Ann Wickham puts it. Broken symmetries are rich in alternatives.

We find this in music too. The symmetries of *Die Kunst der Fuge* are almost miraculous but they are also, in a sense, implacable; it is when Beethoven impatiently breaks away from them that our emotions and imagination are engaged. The symmetries of classical ballet enchant and amaze but it is the departures from them that excite.

As Albrecht Dürer said:

The accord of one thing with another is beautiful. . . . A real harmony linketh together things unlike.

And as H. Osborne recently wrote, also in the context of aesthetics:

The more complex kind of symmetry which consists not so much of repeated elements as of balance and weighting . . . has closer affinities with the aesthetic. . . . Symmetry in this extended sense cannot be subjected to formula or rule but is directly detected by those who have developed the skill of artistic appreciation.

In science we are still developing that "skill of artistic appreciation" and I think that we should take seriously the words of Paul Feyerabend:

If one separates sciences and art . . . if all communication between them is cut, then art will continue to exist but sciences will not.

Grand Unification

Briefly to recapitulate: we have seen the bringing together of the electromagnetic and weak forces in the electroweak unification. This is not, however, a complete unification, because each of the forces, after they have been placed into a definite relationship with each other, is still characterized by its own strength, in the jargon, its own coupling constant, and these coupling constants are related by the parameter, the Weinberg Angle, which is not itself specified in magnitude by the theory; a complete unification would tell us the magnitude of this angle.

We progress by looking for the further unification that would combine the electroweak interaction with the strong quark-gluon interaction, which in our discussions to this point is unrelated to it except insofar as quarks manifest all three interactions. These theories of three-way unification are known in the trade as Grand Unified Theories, or GUTs.

If such Grand Unification of the strong, the weak, and the electromagnetic interactions has indeed been decreed by Nature, then its implications are profound. In particular it would mean that above some energy of unification the distinction between the interactions and the properties of quarks and leptons would fall away; they would interact indifferently to their quark or lepton nature, and consequently, in those interactions, their natures could be interconverted: a quark could be converted into a lepton and vice versa.

The X Particles

These interactions of interconversion would be mediated by some particle or particles, call them collectively the X particles, whose mass corresponds to the energy of unification. For example, two quarks might collide and disappear with the production of an X particle, which might then decay into, say, a positive electron and an antiquark. Such reactions of interconversion would be abundant above the energy of unification where there would, by definition, be enough energy around to make the X particles with ease. But they could also manifest their effects, very rarely, in our own low-energy world because the X particles, however heavy they might be, could be brought into fleeting existence by the embezzlement mechanism to which I have already several times referred.

Nucleon Decay

In fact, the very reaction that I have just described to illustrate the interconversion of quarks and leptons, which will take place freely above the unification energy, must also be possible inside an ordinary nucleon in our ordinary everyday world. Thus two of, say, a proton's complement of three quarks could collide inside the proton, producing the electron and the antiquark; the latter could then combine with the third of the proton's original quarks, which had been a spectator to the X particle's manifestation, to give a neutral π-meson so that the net result would be the decay of a proton into a positive electron and a neutral π-meson. There would be many other additional possible modes of nucleon decay. If Grand Unification obtains in Nature, then there is no natural way of stopping such nucleon decay. Several large-scale experiments are under way in various parts of the world looking for this spontaneous decay of nucleons. No convincing signal has so far been seen and the inferred limit of the lifetime of the process is greater than 10^{32} years.

The rate to be expected for this proton decay will obviously depend on the mass of the X particle because the bigger this mass the shorter the period of time for which it can be brought into fleeting existence by the embezzlement mechanism—in fact the decay rate depends on the inverse fourth power of this mass. Such a possibility is being taken very seriously not only because of its aesthetic attractiveness but also because GUT schemes indeed can be found that bring about a complete unification of the three quantum forces with a single strength, or coupling constant, above the unification energy and are able to predict the value of the Weinberg Angle in very satisfactory agreement with experiment. This agreement determines the associated unification energy: it is very large, of the order of 10^{15} GeV, so that the mass of the X particle should be about 10^{15} times bigger than that of the proton. To make such an X particle using state-of-the-art accelerator technology would need a machine several times bigger than the solar system. We are obviously not going to be able, therefore, to check on the existence of the X particle by making it directly in the laboratory, but as we have seen, if it exists it must mediate proton decay, which accounts for the intense present interest in such searches. The simplest GUT, the so-called SU(5) symmetry scheme, predicts a proton lifetime of only about 10^{30} years, and this now seems to be excluded by the experimental limit that I quoted, but other GUTs can give much longer lifetimes and are by no means excluded.

The Decay of the X and X̄ Particles

Of what use is Grand Unification and its X particles in our search for the solution to the two Big Bang puzzles: the enormous photon-to-nucleon ratio and the absence of antiparticles in today's Universe?

As we pass the Planck Time and emerge into the realm where we can apply physics-as-we-know-it, if still, in part, speculatively, the characteristic temperature corresponds to the Planck Mass of about 10^{19} GeV, which is well above the putative GUT temperature, the mass of the X particles, of about 10^{15} GeV. The GUT symmetry, whatever it is, is unbroken and we have a single interaction, a single force in its equal and equivalent manifestations, that will ultimately be distinguished as strong, weak, and electromagnetic.

This early world, between the Planck Time and the GUT energy, is the realm of the X particles, which are the first recognizable sign of what is to become our eventual Universe. It will remain like this until the Universe has cooled to an energy corresponding to the X particle mass; this takes about 10^{-36} seconds or so. After this time the GUT symmetry begins its spontaneous breaking and its associated GUT Higgs field concomitantly begins to assume a nonzero value; the X particles decay into their various possible product particles more and more rapidly than they can be regenerated out of the residual heat source of the receding Big Bang. By about 10^{-34} seconds the X particles will all have disappeared; the strong interaction will have split away from the still-unified electroweak interaction and will remain forever distinct except for the faint liaison represented by the rare appearance of fleeting X particles in low-energy contexts such as I have already addressed in the case of proton decay. The X particles will have disappeared by decaying into those particles representative of the strong and electroweak interactions, namely, quarks, gluons, leptons, W and Z particles, and photons, of which the Universe will then consist.

How has this helped with our two Big Bang problems? It has helped because of something that I completely left out in this somewhat breathless rush from the 10^{-43} seconds of the Planck Time to the 10^{-34} seconds when it is all over and we are presented with the familiar materials that I have just listed out of which we must construct the Universe as we know it. What I have left out is one of the very problems that I am trying to resolve, namely, the question of antimatter. I earlier stressed that physics knows no way in which, in a single quantum act, we can distinguish between matter and antimatter so that the apparent absence of antimatter in our latter-day Universe constitutes a tremendous challenge to our understanding of the origins of it all. What I have left out are the anti-X particles. That principle of cosmic democracy that I specified as a touchstone of our understanding now demands that after the Planck Time, when we approach our present-day Universe, the X particles must be matched by a precisely equal number of anti-X particles, which we designate as $\bar{X}$ particles.

What happens when particles and their associated antiparticles decay? The rigorous requirement of physics-as-we-understand-it is that they must decay at precisely equal rates. But if they can decay each in several alternative ways, then there is no fundamental requirement that the rate of decay of the X particle in a particular way should be precisely equal to the rate of decay of the anti-X particle in the antiparticular way. The requirement is only that the sum of the rates of all possible modes of decay of the X particle should equal the sum of the corresponding rates of decay of the $\bar{X}$ particle. Thus the decay of the X particle in a way that produces quarks may take place at a rate different from the decay of the $\bar{X}$ particle producing antiquarks, this difference being made up for by a compensatory difference in some other decay mode not of concern to us here. In this way the end product, after 10^{-34} seconds, of the decay of the X particles and the X̄ particles, themselves originally present in precisely equal numbers, could be an excess of quarks over antiquarks.

Disposal of Antimatter

No law of physics would have been violated and, indeed, we are already very familiar with just such a situation in the experimentally studied decay of the K$\wp$-meson, although I do not go into that here. So it is wholly reasonable to hypothesize that after 10^{-34} seconds, when all the X particles and X̄ particles have decayed, and when the temperature is too low for their significant regeneration, we have a slight excess of quarks over antiquarks. Now quarks and antiquarks, particles and their antiparticles, cannot live together and will annihilate each other in various ways, releasing energy as radiation of various forms, in particular, photons and neutrinos. But since, according to our hypothesis, we begin this cosmic Armageddon with a slight excess of quarks over antiquarks, then when the dust has settled, all the antiquarks will have been annihilated by quarks and we shall be left with the slight excess of quarks plus the vast quantity of annihilation products in the particular form of photons and neutrinos. We have disposed of all the antiquarks.

The Generation of Nucleons

The quarks, to begin with, remain as such, but as the temperature falls further they combine with each other to give the great menagerie of familiar strongly interacting nuclear particles, including the mesons, which, although they contain antiquarks, can be formed out of the still-considerable energy that the temperature represents. This assembling of the familiar nuclear particles takes about a microsecond; after about 10 microseconds the only particles remaining are nucleons and π-mesons, which are the lightest of the mesons and therefore the ones that can be sustained in significant abundance in the lowest temperatures, i.e., for the longest time; after about a millisecond the π-mesons have also all gone, the temperature now being too low to sustain their continued regeneration, and, with just neutrons and protons as nuclear consistuents to work with, we enter the epoch of light element synthesis, which I have already discussed.

The Resolution of the Two Big Bang Puzzles

I have left things out: I have left out the fact that in the course of the story above, we have passed through the temperature at which the electroweak symmetry breaks and the W particles and Z particles disappear from the scene; this is after about 10^{-10} seconds when the temperature has fallen to about 100 GeV, the mass of the two species of particle in question; so at about 10^{-10} seconds the weak interaction splits away from the electromagnetic and they become, and remain, distinct in their manifestations. (I should mention, in passing, that this electroweak symmetry breaking might itself be a generator of the matter/antimatter asymmetry, although I am telling my story just in terms of the X and $\bar{\text{X}}$ particles.) I have also left out my emphasis on the purpose of constructing the story in the first place: to resolve the two Big Bang puzzles. This we have, in fact, done at one blow because we have seen that the world, after a small fraction of a second, consists just of matter since the antiquarks, which would have been the building stuff of antimatter, have been annihilated and replaced by energy in the form of radiation.

All we now need to do is to suppose further that the excess of quarks over antiquarks after the 10^{-34} seconds of the existence of the X particles and the $\bar{\text{X}}$ particles was about 1 part in 10^9 so that following the Armageddon of mutual quark/antiquark annihilation we would be left not only with a world of just matter without antimatter but also with a world in which radiation, the product of the quark/antiquark annihilation, dominated matter by the observed factor of 10^9. At this stage our understanding why that factor should be 10^9 rather than 10^5 or 10^{15} remains without specific derivation, because we have not yet uniquely defined the GUT that lies behind it, but it is a very reasonable factor in the light of present understanding and expectations.

Remaining Big Bang Problems

Are we, therefore, content to leave the Big Bang, happy that it has given us the makings of our Universe as we see it? By no manner of means. We have understood the recession of the galaxies, the Expanding Universe. We have understood the 3°K microwave background radiation. We have understood the origin of the light elements. We have understood that our Universe is made of matter. We have understood the extraordinary brightness of our Universe as expressed in the 10^9 photon-to-nucleon ratio.

Let us now examine the things that we have not, so far in my story, understood. I will call these: (1) Why is the Universe so smooth? (2) Why is the Universe so lumpy? (3) Why is the Universe so flat? (4) Where have all the monopoles gone? (5) What about the cosmological constant?

Why Is the Universe So Smooth? I have already referred to the remarkable isotropy of the microwave background radiation: if one looks out to the farthest stretches of the visible Universe in one direction and then in the diametrically opposite direction, the microwave background radiation has the same intensity, in those two directions, to much better than one part in a thousand. Why is this a worry? Is it not just what we should have expected if everything started out isotropically from a Big Bang? Naively yes, but on second thoughts, no. The reason for the change of mind comes from consideration of the way in which the microwave background radiation got loose and flooded space. Do not forget that, at any epoch, the observable Universe, which is all that we are allowed to talk about, the domain within which communication can occur, is limited by its age.

So if we wish to understand the very precise equality of intensity of the microwave background radiation now coming to us from opposite sides of our present-day Universe we must go back to the time when that radiation flooded out and make sure that there was, at that epoch, a common site for that flooding out so that at that time the radiation we now find coming to us from opposite directions was all mixed up with itself and would naturally be apportioned equally among those bits of space that are now so far apart.

What do I mean by "flooding out"? The point is that electromagnetic radiation interacts strongly with free electrons and can scarcely struggle through a medium in which they are abundant. But electrons would remain free in the early Universe until the temperature had fallen to such a degree that they would be incorporated into electrically neutral atoms and no longer be able to interfere with the passage of electromagnetic radiation. We are asking the question of how long it took for the Universe to cool to a temperature corresponding to typical binding energies of electrons into atoms. The answer is about 100,000 years. So for about 100,000 years electromagnetic radiation was pretty well tied up wherever it started from and only after that could it flood out unimpeded through all space and so communicate with other regions of space. At that epoch the then-visible Universes now corresponding to the two opposite directions that we consider in space today, from which the microwave emissions are so extraordinarily equal, were separated from each other by about a hundred times their own dimensions. It seems, therefore, that the two sources of the microwave background radiation that are now equal in intensity to much better than one part in a thousand have never been in causal contact, have never had the opportunity of comparing notes with each other, have never had available to them a mechanism for arranging that their intensities should now be so remarkably equal. This is the first major problem that remains to be resolved before the Big Bang can be said to approach complete sense.

Why is the Universe so Lumpy? Although, as I have just stressed, the Universe is incredibly smooth from the point of view of the distribution within it of the microwave background radiation, the fossil relic of the Big Bang itself, it is certainly far from smooth on the smaller scale of the local distribution of matter that we see as stars and galaxies: they, and we ourselves, are obviously lumpinesses of some local importance.

Averaged out over sufficiently large regions of space the Universe becomes smooth again: the lumpiness is the same everywhere. But can we understand the local clumping of matter into stars and galaxies that gives to the Universe its essential texture? In a qualitative sense there is no problem because the force of gravity will obviously lead to the congregation of matter into blobs of various sorts, the smaller ones stars, the larger ones the collections of stars that we call galaxies. Fluctuations of density shortly after the Big Bang will provide a kind of initial seeding of the Universe that will then be built upon by the inexorable force of gravity as the Universe ages. We can easily address the question of what must have been the scale and degree of those initial density fluctuations in order to constitute the sort of seeding that would have led to a Universe of today's lumpiness.

The answer is the very uncomfortable one that just after the Planck Time, when the materials that led eventually to today's Universe first presented themselves, matter must have been distributed much more smoothly than that of a normal gas in thermodynamic equilibrium. In other words the ordinary fluctuations of density associated with random, uncorrelated motions would have constituted a seeding that would have led to a much coarser-grained Universe than the one we see today. An initial uncorrelated randomness is, however, what we should unavoidably expect: it is difficult to see how a smoother initial distribution than that associated with random motion could have been brought about other than by some sort of "deliberate" placing of the elementary units of matter, whatever they were at the time, in a kind of semiordered array of just the correct graininess, or by some sort of stretching-out process that would dilute the randomness of the initial distribution. This is our second major residual Big Bang problem.

Why Is the Universe So Flat? The phenomenology behind this question concerns the matter and energy density of the Universe. The Universe is today expanding, as evidenced by the mutual recession of the galaxies. But the mutual gravitational effect of the matter that the Universe contains is tending to make the Universe collapse back on itself: the galaxies as they move apart are all the time exerting gravitational forces upon each other that are tending to bring them back together again. What is the balance of these effects? Does the Universe contain enough gravitating stuff in the form of matter, plus other forms of energy, with its $E = mc^2$ mass equivalence, to halt eventually and reverse its present expansion and bring everything back together again into a Big Crunch? Or is the density sufficiently low for this not to happen and for the Universe to continue expanding forever?

We can define a closure density for the Universe such that the expansion and the gravitational attraction just balance, such that the expansion would continue forever but more and more slowly as time went on, never quite reaching the turnaround point. Call this closure density unity and measure the actual density, Ω, in terms of it. What should we, offhand, expect for Ω? How much matter should we have expected the Big Bang to have produced?

We surely have not the faintest idea. Why should not Ω have been a million? Or a millionth? Certainly for it to have been anywhere near unity would have been either an unbelievable coincidence or a very significant clue about the nature of creation. What, in fact, do we know about Ω?

The Density of the Universe. We can get a good lower limit to Ω simply by looking at the shining matter in the sky: weighing and counting the stars. We believe that we have pretty good theories of stellar evolution and have a good idea of the mass of a star from its brightness and spectral characteristics. We also have a good idea of the structure of galaxies in terms of the stars that they comprise. So we think we can estimate the amount of matter in the bright lights of the Universe with reasonable accuracy: it corresponds to $\Omega \sim 0.01$ or so and is obviously a lower limit to the actual value of Ω.

When I was describing the synthesis of the light elements in the Big Bang I said that the predictions about their cosmic abundance depended on today's mean density of nucleons in the Universe, rather sensitively so in some cases. The value of Ω needed to gain good agreement in all cases where relevant data are available (^{2}H, ^{3}He, ^{4}He, and ^{7}Li) is $\Omega \sim 0.1$. This value is consistent with the $\Omega \sim 0.01$ of the shining matter; comparison of the two values implies that there may now be a lot of nonshining nucleons around, perhaps in the form of undetected dust or small objects up to about the size of Jupiter, comets, asteroids, bricks, and so on; above about the size of Jupiter matter begins to glow as the size and condition of small stars is approached; we have, in fact, only very poor knowledge of what the number of such small "brown" stars might be.

The evidence from the abundance of the light elements relates to what were nucleons at the time of the synthesis of the lightest elements, a couple of minutes after the Big Bang; after that time a lot may have disappeared into black holes and have lost their individuality, and, indeed, their identity, in the faceless, singularity that I discussed earlier, but in their disappearance they would have added to the gravitating mass of the black hole whatever of their own mass had not been radiated away as they fell in.

Missing Mass. This tentative conclusion that the Universe may contain a lot of nonshining gravitating stuff is supported by detailed studies of intragalactic and intergalactic dynamics. For example, one can study the orbital speed of stars as they circulate about the center of their galaxy. If the only gravitating mass in the galaxy were that represented by the stars that one sees shining within it, then the orbital velocity of stars toward the edge of the galaxy would fall off with their distance from the center of the galaxy in a well-defined way. Such a falloff is not found; the oribital velocity remains approximately constant. This can be understood only if the galaxy has some sort of large, invisible halo of gravitating matter that keeps up the pull on the more distant stars. Similarly, one can study the motions of galaxies relative to each other in clusters of galaxies. Again, those motions are faster than can be understood simply on the basis of the gravitational forces exerted by the shining matter. The inference from these studies is that this nonshining "missing mass" is enough to bring Ω up to perhaps 0.1 to 0.2; some studies, involving infrared satellite observations, suggest even higher figures, toward unity, for Ω. Studies on the number density of galaxies as a function of distance out toward the edge of the visible universe have yielded $\Omega \sim 1$. At this stage we might ask whether a value of exactly unity would be acceptable for Ω. It seems that this would not be so if the chief component of Ω were nucleons whether still such or now disappeared into black holes because this would wreck the beautiful agreement between the cosmic abundances of the light elements and the previsions of the Big Bang scenario for their production, with respect to which, values of Ω significantly greater than 0.1 cannot be tolerated. Attempts have been made to reconcile the light element abundances with $\Omega \sim 1$ by invoking a fractionation and nonuniform distribution of neutrons and protons at the epoch of nucleosynthesis such as might be due to the difference of their electrical charges, but these attempts encounter formidable difficulties that have not been overcome.

At the moment the best bet seems to be that the contribution of nucleons, past or present, to Ω is about 0.1 and that if $\Omega = 1$ were to be correct, then the rest of the matter would have to be in some nonnucleonic form; furthermore, that nonnucleonic matter must not clump with ordinary matter in galaxies and galactic clusters but be more uniformly spread out throughout space where its gravitational effects would not be significantly manifested on the shining matter to which our direct dynamic observations must chiefly relate.

The Astonishing Value of Ω. Leaving out of account for the moment the possibility that Ω might be exactly unity, what are we to make of the fact that it is at least 0.1 and, as a very generous upper limit on the basis of the dynamic studies, less than 2? I have already indicated how very surprising it is that Ω should have a value anywhere near unity on general a priori grounds. But a value today in the range 0.1 to 2 is even more astonishing if one works back in time.

Because of the general-relativistic nature of the expansion of the Universe, a value of Ω in the region of unity is unstable: to be within the range 0.1 to 2 today, Ω would have had to have been within one part in 10^{15} of unity at the time when the light elements began to form and within one part in 10^{49} of unity at the time of the GUT symmetry breaking when the strong force split away from the electroweak interaction. Such fine-tuning becomes unbelievable, and one feels that some explicit mechanism must have been operating that would make Ω exactly unity or indefinitely close to it.

I will look in a moment at how we might have $\Omega = 1$ if we do not want a contribution of more than 0.1 or so from nucleons but will now pause to tie in what I have just been saying with the name of the problem: Why is the Universe so flat?

Flatness. Flatness is a technical term obviously related to the curvature of space, which is due, as I have indicated in general terms, to the matter that it contains: the more and denser the matter, the greater the curvature. If $\Omega > 1$, everything will eventually collapse back in a Big Crunch; in terms of the curvature of space this is because space would then be closed and curve back on itself much as the lines of longitude starting from the North pole curve back on themselves and meet again at the South Pole: a mini Big Bang at the North Pole that shot spacecraft off horizontally skimming the surface of the earth with orbital velocities along the lines of longitude would be followed by a mini Big Crunch as the spacecraft met again at the South Pole. If $\Omega < 1$, space is again curved but is not closed: such curved space describes an open Universe. If $\Omega = 1$, the Universe is flat and is described by the familiar Euclidean geometry that we can picture through the rectilinear Cartesian system of an ordinary piece of graph paper, as against the curved straight lines of longitude on the surface of a globe that we should have for $\Omega > 1$ or the diverging curved straight lines like a pincushion for $\Omega < 1$.

Although I have now exposed, somewhat in caricature, the technical meaning of flatness, and the relationship of Ω to it, I will continue to discuss Ω as such and to ask whether we might understand how Ω could be exactly unity, if some mechanism were found for making it so, despite our reluctance to make a contribution of more than about 10% of it from nucleons.

Dark Matter: Photons. Remember, to begin with, that energy in any form gravitates because of $E = mc^2$. In particular electromagnetic radiation gravitates: light trapped inside a cubical box made of inward-facing mirrors could be weighed on a balance. We understand this because, loosely speaking, as the photons bounce up and down between the top mirror and the bottom mirror they lose and gain energy from the gravitational field so that they are redder at the top and bluer at the bottom and therefore give more of a downward push to the box on bouncing off the bottom mirror than they give an upward push on bouncing back off the top mirror. We might then think that the microwave background radiation, which is so overwhelmingly dominant in terms of the number of particles of energy that it represents relative to the number of nucleons in the Universe, by that factor of 10^9, could make a significant contribution to Ω. This is not, however, so, because the energy associated with each microwave photon is so small and the total contribution to Ω is less than 0.01%. Similarly all other forms of electromagnetic energy add up to a negligible amount.

Dark Matter: Neutrinos. There is, however, one other form of radiation that must be very abundant even though we have no direct evidence for it: neutrinos. In that cosmic Armageddon of quark/antiquark annihilation that I described as following the decay of the X and $\bar{X}$ particles, about 10^{-34} seconds after the Big Bang, during which time what are today the microwave background photons were generated, there must also have been generated, in roughly equal abundance to the photons, neutrinos of all species. These must still be around. As with the photons, their energies will have been diminished in the subsequent overall expansion of the Universe, but if they, unlike the photons, have finite intrinsic mass, that will not have been diminished. We do not know what mass, if any, neutrinos possess; it is often supposed to be zero. There is, however, no fundamental reason that we know why this should be so; a finite mass is eminently possible. Indeed, Soviet experiments persist in claiming a mass of about 30 eV for the neutrino that is associated with the ordinary electron although experiments elsewhere have critically questioned this. If neutrinos were to have a mass of, say, 10 eV, that is to say, a mass about 10^8 times less than that of the proton, and if they were to be 10^9 times more abundant than nucleons, as the microwave background photons are, and we should indeed expect rough equality of numbers of the neutrinos and of the microwave photons, then their contribution to Ω would be 10 times greater than $\Omega \sim 0.1$ of the nucleons, namely, about unity, and we could have achieved $\Omega = 1$. Such hot dark matter would indeed not clump in galaxies and so would not make much of a contribution to the dynamic determination of Ω based on observations of the motions of shining matter.

In all such speculative cases, however, one must ask what other consequences there might be. A key such question is whether there would be consequences for the nature and distribution of galaxies, whose formation will obviously be affected by whatever forms of energy and matter were around at the critical time of their inception as fluctuations in the primordial matter density. In the case of massy neutrinos it seems that they would have led to a much more pronounced clumping together of galaxies than we find in practice and this argues against their candidacy as the missing mass.

If, however, there were other overridingly powerful agencies at work bearing on galaxy formation, then this conclusion would not necessarily follow. For example, it is now becoming clear that in addition to the well-recognized gathering together of galaxies into clusters and superclusters, there is a tendency of galaxies, galactic clusters, and superclusters to distribute themselves on a much grander scale in great sheets and chains with voids in between almost like a sponge or foam. Such a de facto structure might let neutrinos back into contention as missing mass candidates, but it is too early to say more.

Dark Matter: Other Candidates. Particle physics is alive with alternative candidates to provide the dark missing mass. None of these is an as-yet-known particle, but their possible existence is suggested by various needs that our present understanding, or lack of understanding, of particle physics has brought forward. For example, there is the axion, which would be a particle of indeterminate, but probably very small, mass whose existence is suggested to safeguard the respecting of certain very well established symmetries that Nature displays in the strong interaction.

Supersymmetry. A whole set of possible particles, some of which could be candidates for the missing mass, is associated with the idea of supersymmetry. The basic idea here is that every particle should be associated with a particle belonging to the opposite symmetry family: every ordinary particle of half odd-integral intrinsic spin would be associated with a supersymmetric partner of integral spin and vice versa. Thus the ordinary photon of spin 1 would be associated with a "photino" of spin ½ and so on.

The Mass Hierarchy Problem. The chief reason for believing that such an organization might in fact obtain has to do with the problem of the masses of particles. We do not understand why most of the particles of our experience have the masses they do. Particle masses range from zero for the photon through very small masses, possibly also zero, for the neutrinos through 0.5 MeV for the electron, 1 GeV for the proton, 100 Gev for the W and Z particles, and upward, although there are as yet no other well-established stopping-off points, toward the hypothetical GUT mass of perhaps 10^{15} GeV and the Planck Mass of 10^{19} GeV.

We do not understand how this tremendously broad mass spectrum gets set up in the first place nor, and this is relevant for our present consideration, do we know what stabilizes it when once it has been set up. The point of the last remark is that all particles must affect each other because their force fields can bring one another into fleeting, so-called virtual, being, as we have seen in discussing quantum electrodynamics. This will affect the masses that the particles display in practice because one can never have a particle unattended by such a virtual entourage whose various interactions contribute energy, and therefore mass, to the system. When this idea is pressed through, it seems impossible to understand why all particle masses do not get raised by such mechanisms right up to the Planck Mass, which is patently not the case. This is known in the trade as the mass hierarchy problem.

If, however, every particle were to possess a supersymmetric partner, then the situation could be resolved and the masses stabilized because it is a general result that the contribution to the mass of a particle by its virtual attendants has opposite sign, but the same magnitude, for particle and superpartner particle attendants. So if particles and their associated superpartners had the same masses, then the mass of the particle that they attended would not be shifted by that attention. If the particle and superparticle masses are not the same, then there will be a resultant mass shift, but it will be limited. Clearly, superparticle masses, if such particles indeed exist, are larger than the masses of the associated ordinary particles because the superpartners have not shown up yet, but that does not mean that they do not exist. Searches for this supersymmetry world will be a chief objective of the next generation of accelerators, but thus far we have no direct experimental indication of their possible existence.

If, indeed, superparticles exist, then certain of them would become possible candidates for the missing mass. I will return shortly to discuss why I have gone on for so long about the possibility that Ω might be exactly unity when all that was necessary to define the astonishing nature of this third in the list of problems facing the Big Bang was the experimental observation that it is somewhere in the range 0.1 to 2.

Where Have All the Monopoles Gone? First: what is a monopole? In our everyday world we are familiar with isolated electric charges such as those with which the electron and the proton are endowed and we are also familiar with magnets, which we describe as dipoles and speak of as having a north pole at one end and a south pole at the other; but we have never seen an isolated north or south pole: electromagnetism is asymmetric as between electricity and magnetism. There is no fundamental reason for this asymmetry and the possibility of the existence of isolated magnetic poles —magnetic monopoles—has been discussed off and on for a long time.

GUTs in fact demand the existence of monopoles and offer explicit accounts of their structure: their mass is about 10^{16} GeV, which means that we cannot make them in the laboratory since that is 10 times the mass of the X particle, which we saw would require an accelerator larger than the solar system for us to bring into being; they have a core perhaps 10^{-25} cm or so across, within which Grand Unification reigns and which holds most of the mass; just outside this core X and $\bar{X}$ particles are numerous, and as one moves out to the edge of the monopole, at about 10^{-15} cm from its center, one encounters a shell of W and Z particles. Such monopoles would have striking properties: for example, a proton meeting a monopole would encounter the X particles that we have seen are involved in the putative decay of protons into, in the example I gave, positive electrons and neutral π-mesons so that such an encounter between a proton and a monopole would very probably result in the catalyzed decay of the proton; the monopole would go on its way destroying all nucleons in its path, and that would be bad news.

It is obvious that there are not many monopoles around today. One can set a limit to the number of monopoles by the observation that galaxies, at least our own, the Milky Way, possess a small overall magnetic field of about a microgauss. Now monopoles would be accelerated in a magnetic field and so would drain the galactic magnetic field away as they removed energy from it in this way. But the magnetic field is there and this sets quite a low limit, the Parker limit, to the abundance of monopoles.

What have monopoles to do with the Big Bang and why are

they a worry for it? This has to do with the spontaneous breaking of the GUT symmetry that will happen when the Universe, after 10^{-35} seconds or so, is cooling through the temperature below which the thermal energy cannot sustain the existence of the X particles. The symmetry then breaks and the GUT Higgs field takes its nonzero value as I described earlier. But, as also described, the sense of the symmetry breaking is not predetermined; its nature and the properties of the resultant volume of space that participates in that particular act of symmetry breaking and that grows as a kind of bubble with the velocity of light are parameterized by the associated phase angle of the emergent GUT Higgs field.

Now different regions of space will break their symmetry at slightly different times, because of the tendency to supercool that I mentioned earlier, so that these different regions will run into each other in a random way, and at their boundaries there will be conflict between the different phase angles, the different senses of the symmetry breaking. This is very much like the situation that arises when supercooled water freezes and the little crystallites run together with boundaries between them at which the different orientations of the crystal growth meet and constitute crystal defects. In the case of the spontaneous breaking of the GUT symmetry, when two regions of differently broken symmetry, characterized by different GUT Higgs field phase angles, run together, we get what is called a domain wall, which is very massive; when three regions meet we get a string, to which I shall refer later; when four regions meet at a point we get a point defect, which is the monopole—in the jargon, a monopole is a topologically stable knot in the GUT Higgs field.

Now the sizes of the broken symmetry domains within which there is a single phase angle, that is to say, a unique sense of the breaking of the symmetry, cannot be much bigger than the size of the observable Universe in that region at that time; we can therefore easily estimate a lower limit to the number of monopoles that must have been formed. The number is embarrassingly large: the monopoles would dominate the energy density in the Universe and so speed up its evolution so that, for example, the present-day microwave background radiation temperature of 3°K would have been reached after only about a few thousand years instead of the observed fifteen thousand million years, and shortly after that the Universe would have collapsed back to a Big Crunch. The monopoles clearly must be got rid of if the Big Bang is to survive. This is the fourth residual Big Bang problem.

What about the Cosmological Constant? This problem is thought by many to be the most intractable facing us today; it can be stated very briefly. As I mentioned earlier, the equations of general relativity admit of the introduction of a constant whose dimensions are, in a standard representation, the inverse square of length, that operates together with the mass and energy density in determining the rate of expansion of the Universe. Since the equations permit of its introduction it cannot be kept out without a very good and explicit reason (on the very generally valid principle in physics that everything that is not forbidden is compulsory), and all that remains is to discuss its likely magnitude. And since it is a constant it must always have been the same. Now at the beginning of physics-as-we-know-it, at the Planck Time, the only measure of length available was the Planck Length of about 10^{-33} cm, so we should offhand expect the magnitude of the cosmological constant to be about 10^{66} cm^{-2}. Our observations on the expansion of the Universe, in particular, the limit that can be set on the rate of change of the rate of expansion, tell us that in fact the cosmological constant is less than 10^{-55} cm^{-2}, so that we have a discrepancy between our expectation and experiment of a factor of 10^{121}. This discrepancy is so large that it almost defies comment. Where has the cosmological constant gone to?

The issue is made the more perplexing by the fact that spontaneous symmetry breaking, as I mentioned in introducing the properties of the Higgs field, implies that there is a kind of explosive potential associated with it; this can be parameterized in the form of a contribution to the cosmological constant so that the cosmological constant as we determine it today, or rather as we do not determine it, must be made up of the ur-constant associated with the emergence of the Universe from the Planck Time plus whatever effective cosmological constants must be associated with the subsequent symmetry breakings down to the present time. This would not compound the issue if the latter contributions were expected to be small, but they are not. The effective cosmological constant associated with the last of the symmetry breakings, that of the electroweak symmetry at about 100 GeV or 10^{-10} seconds after the Big Bang, contributes about 10^{-4} cm^{-2}, which is, of itself, about 10^{51} times larger than the experimental limit, and we should certainly expect that that associated with the much higher energy GUT symmetry breaking at about 10^{15} GeV or 10^{-35} seconds would be much larger.

It is evident that the trick necessary to enable us to understand that the sum of these various contributions should be so unimaginably tinier than any of them separately is one that constitutes a major challenge for us to turn. It may well be the case that we cannot understand it at all except by saying that it is a law of Nature, which is what we always do if things get desperate enough.

This has completed my brief survey of problems that must be addressed if we wish to consider the Big Bang model as acceptably founded.

Inflation

The key to the resolution of most of the problems I have just set out appears to be in the concept of inflation. This is that at some very early stage after the Big Bang, around the time I have earlier given as that at which the GUT symmetry begins to break, the

Universe entered a phase of exponential expansion of its size that continued until it was inflated by an enormous factor of perhaps 10^{50}, all this taking perhaps 10^{-32} seconds.

Why, in Heaven's name, should this extraordinary thing come about? The simplest way to see that something of this kind is not only reasonable but almost unavoidable is through the original inflationary model of Alan Guth, from which the rough numbers above are taken, although this specific scenario is now abandoned. Start from the very early Universe at a very high temperature, after the Planck Time but above the GUT transition temperature, with the GUT symmetry unbroken and with the associated GUT Higgs field therefore at a value of zero. The Universe now cools below the transition temperature but supercools, as I have earlier suggested might be likely. The lowest energy would be with broken symmetry and a finite value for the GUT Higgs field, but the supercooled state, with the GUT Higgs field zero, has higher energy, which is, as it were, locked up in the GUT Higgs field. This extra energy is very considerable—in a typical GUT it is the $E = mc^2$ equivalent of about 10^{73} gm/cm^3 or roughly the mass energy of a large star compressed into the size of a proton—and it behaves, in effect, like a cosmological constant and brings about the rapid exponential expansion that is inflation. The Universe inflates at a speed greater than that of light although this causes no anxieties in the context of relativity because it is just a change of the overall dimensions of the Universe without any transmission of information between the parts.

During the inflation the temperature, because of the supercooling, drops below what was expected on the noninflationary scenario that I sketched earlier and the rest of the scenario is temporarily put into abeyance. However, eventually, after perhaps 10^{-32} seconds, the GUT symmetry completely breaks, and the release of energy raises the temperature to what it would have been on the earlier scenario, which then takes over as before.

We therefore recover all the results and successes of the earlier version of the Big Bang but with the great difference that between the Planck Time and the onset of history the Universe has been torn apart in this vast inflation by a factor of 10^{50} or so, which has dramatic consequences for the problems that I have presented as being left over by the earlier scenario.

A Variety of Inflations

Before returning to the problems of the Big Bang and indicating the impact of inflation upon them I should enter a brief caveat or disclaimer. That is that there is no presently agreed scenario for inflation. The one I have just sketched is only the first of many that have been proposed. Some versions of inflation, as in Guth's original, involve a rapid transition from the supercooled to the normal state, namely, from the zero value for the GUT Higgs field to the finite value corresponding to the broken symmetry. Such scenarios fail because they lead to final-state Universes that are dominated by a single region, a single bubble, of the broken symmetry with lesser bubbles clustered around it as the result of the process of random nucleation; although the individual bubbles do not coalesce, energy would be concentrated near the surface of the largest bubble without any means of securing a more nearly uniform distribution; this is nothing like the Universe we see.

Other scenarios, such as one due to Andrei Linde, overcome this problem in a somewhat *ad hoc* and fine-tuned way by hypothesizing a relationship between the energy of the GUT Higgs field and its value such as would lead to a slow, rather than a rapid, transition from the symmetrical to the broken symmetry state. In this case the broken-symmetry phase forms domains that run into each other in the course of their formation, each continuing to inflate until the symmetry breaking is complete when a single domain is large enough to encompass what is now our entire present-day Universe. This scenario also encounters problems whose overcoming enable it to explain satisfactorily why our Universe is much more nearly uniform in its gross structure than predicted by the earlier scenario, but it still leads us to expect a much greater variability of the microwave background radiation than is observed.

Further scenarios, such as the chaotic inflationary model, also due to Linde, do not involve symmetry-breaking transitions or supercooling at all, with their unusual relationship between energy and field that I have been discussing so far for the GUT Higgs field, but just invoke the intervention of some scalar field, "scalar" being used in the sense that I defined it before, of unspecified origin, whose associated energy goes as some reasonable power of the field in a more familiar way. The random quantum fluctuations of this scalar field, which we can think of as coming about by courtesy of the usual embezzlement mechanism, could lead to occasional regions of high field, whose energy would behave like an inflationary cosmological constant, causing the now-familiar expansion. Such models can lead to reasonable fluctuations in the intensity of the microwave background radiation that do not conflict with observation. (I might remark in passing that it has been suggested by A. Vilenkin that even closed universes will be kept from collapsing by the quantum fluctuations of the scalar field responsible for inflation.)

What I am getting at is that there are many ways in which inflation might have occurred; in fact, it is rather like the cosmological constant itself: once thought of it is difficult to get rid of it even if you want to (cf. the aborigine boy's old boomerang).

Inflation as Cure for the Big Bang Problems

I have not yet said why inflation is a good thing with respect to the resolution of the residual Big Bang problems.

Why is the Universe so smooth? Postinflation there is no problem about lack of early causal connection between remote regions of our present Universe because our present Universe has evolved from a size 10^{50} times smaller than it was thought to have been in the preinflationary model of the Big Bang; that then tiny region was in complete thermodynamic equilibrium, and the regions, now far separated, of our present Universe from which we see the microwave background radiation coming with remarkably equal intensity were completely overlapping.

Why is the Universe so lumpy? This remains a tricky problem because the details of the predictions depend on the details of the models and are not given just by the qualitative notion of inflation. However, some inflationary models, although not the simplest, do lead to a spectrum of inhomogeneities that resemble that of observed galactic distribution. There is certainly no necessary conflict with the notion of inflation of the slower kind.

Why is the Universe so flat? This problem is also given an immediate explanation by inflation. I earlier remarked that $\Omega = 1$ corresponds to flat or Euclidean space rather like rectilinear graph paper, whereas $\Omega > 1$ corresponds to curved space like the lines of latitude and longitude on the surface of a sphere and that $\Omega < 1$ is like a pincushion. But the effect of inflation would be like blowing up the sphere or the pincushion to a gigantic size but continuing to look just at a little part of it because our visible Universe is just a minute fraction of the Universe before inflation. But if we blow up the sphere or the pincushion to a gigantic size, the originally curved lines upon them will be stretched out into an accurately rectilinear grid and we shall be looking at flat space. Thus inflation automatically generates flat space, i.e., $\Omega = 1$ to a very high degree from no matter what starting point.

This is why I went on at some length about the possibility of there being missing mass to top up Ω to exactly unity even though we were not keen to admit a contribution of more than $\Omega = 0.1$ or so from nucleons past or present. I should remark here that some theories of quantum gravity seem, without invoking inflation, automatically to generate a Universe with $\Omega = 1$, but this is not to say that inflation does not also occur.

Where have all the monopoles gone? The initial production of monopoles is different according to the different inflationary models since the generation of the monopoles is due to the running together of different regions of differently broken GUT symmetry and the degree to which this happens depends on whether the breaking is quick or slow and so on. However, the inflation itself vastly dilutes the monopole density and we have no difficulty in understanding their very low abundance today.

What about the cosmological constant? Inflation, of itself, has nothing to say about the fundamental cosmological constant that is defined at the end of the Planck Time before the present GUT-based considerations begin. As we have seen, the energy locked up in the GUT Higgs field after symmetry breaking on those models that invoke this, or in the *ad hoc* scalar field for other models, behaves like a cosmological constant, but that is another story that I have just told. So the problem of the cosmological constant remains for further discussion; I shall return to it.

Cosmic Walls and Cosmic Strings

Before I leave my sketch of the early Universe I will just touch upon a few matters that naturally suggest themselves from what I have said so far. One concerns the running together of the regions of differently broken GUT symmetry. I have discussed the magnetic monopoles that would be formed by several regions coming together at a point. I have also mentioned the domain walls that would separate two regions over an area, and there can obviously be lines or strings where three regions come together.

These strings and walls would be somewhat like one- and two-dimensional monopoles; they would be massive and have possible effects on the galactic scale. There is certainly no dramatic evidence for them such as a partitioning off of part of the Universe by a gravitating wall stretched across it. It is possible, however, that we are seeing some fossil sign of the influence of cosmic strings and walls in the mysterious chains and sheets into which galaxies and galactic clusters seem to have formed themselves in those sponge-like structures to which I earlier referred.

Unification of Gravity and GUTs

Another matter to which I have referred obliquely is the possible eventual unification of gravity with the other three forces. For this to come about, gravity must itself be given a quantum form. We can, by following standard procedures, tack classical general relativity onto a quantum mechanical GUT, but then the GUT, by itself renormalizable in the sense that I described earlier as disposing in a consistent manner of infinities in its predictions, seems inescapably to acquire new nonrenormalizable infinities. This appears to be an unavoidable stumbling block for all theories involving point fundamental particles like the standard GUTs. We might then hope to do better by first defining a quantum gravity followed by its unification with a GUT, and there has been some, but not compelling, success in the first stage of such a program. Or we might hope to do it all in one go, defining some theory that would of itself contain both quantum gravity and a GUT.

Superstrings

An idea along the latter line that is now being energetically pursued is that of the superstring, which, in its modern form, beginning in 1974, is chiefly associated with Michael Green, John Schwarz, and Ed Witten, although it is now a vast international intellectual industry. "String" here has nothing to do with the cosmic strings that I have just mentioned; it refers to the fact that the fundamental entities that the theory describes are not points, as in conventional theories of electrons, quarks, and so on, but are tiny essentially one-dimensional objects, strings, perhaps 10^{-33} cm long (this dimension being chosen so as to accommodate gravity with the correct strength) either open with two ends or joined in a loop. The various particle spectra and families are generated by various forms of excitation of the strings somewhat like the harmonics of a stretched wire. The strings can also join and link themselves in various ways to describe the interaction between particles and the production of particles.

"Super" in "superstrings" refers to the fact that the theories so constructed have a supersymmetric character, in the sense that I defined before, that emerges naturally. (It is not thought to be possible to construct a nonsupersymmetric string theory.) Superstrings are quantum theories, as we have seen to be essential, but the fact that they are based on one-dimensional entities rather than zero-dimensional points holds out the hope that they will not be plagued, or not so seriously plagued, by the infinities that we saw to arise in the point-based theories and that had to be spirited away by the artifice of renormalization, which merely sweeps them under a neat and self-consistent carpet. Another attractive feature is that superstrings can carry a natural chirality, which is, loosely speaking, the tendency for particles to have a handedness such as we have noticed from time to time. But a most striking thing about superstrings is that they also naturally contain massless particles of intrinsic spin 2 such as can be associated with the graviton, the putative quantum messenger particle of the gravitational force, which must have spin 2, as has been known for decades, and that they can also accommodate the equations of general relativity as a classical limit.

(It is interesting to trace the development of our understanding of gravity: Newton's work established that gravity followed an inverse-square law of force but did not explain why; Einstein's general relativity explained why the law was an inverse square but did not explain why gravity existed; superstrings *require* the existence of gravity.)

Another most important and attractive feature of superstrings is that one can find formulations for them such that after the global superstring symmetry is broken at the Planck Time, what then splits away from gravity can be an acceptable candidate for the sort of GUT of which I have already spoken at length in discussing the post-Planck-Time epoch of the evolution of the Universe; from then on it is as before.

It is even to be hoped, and explicit claims have already been made to this effect, that superstrings may explain where the cosmological constant has gone to. Indeed, by many, superstrings are already being hailed as the theory of everything. It must, however, also be said that by others superstrings are regarded as nothing more than a fascinating mathematical toy. Although it is not yet clear that the nag of the infinities can be completely disposed of, progress is such that surely superstrings must be carrying us closer to the heart of things than we have been before, always bearing in mind, of course, my underlying theme in this book, namely, that we may be inventing, not discovering, and that the only heart to be accessed may be our own.

Superstrings do not define themselves uniquely and several versions are on the market. Acceptable versions, in the sense of accommodating what we know to be so and not accommodating at the same time other things that we know not to be so, are rather few. But all have a very striking feature: they operate in many dimensions.

Many Dimensions

The world of our own experience appears to have three dimensions of space and one of time although we have no fundamental idea of why it is so limited—I shall return to that question later. But superstrings will not work in such a restricted space; technically acceptable versions demand ten or twenty-six dimensions instead of the observed four. That means that in the energy regime of full unification, tucked inside the Planck Time, the world has these extra dimensions and behaves in a way that we cannot describe in our ordinary language because we do not command the necessary imagery. In order to break through to physics-as-we-know-it we must dispose of those extra, unwanted, dimensions in a respectable manner that does not do violence to our four-dimensional world picture of the post-Planck epoch. The mechanism of disposal is called compactification and consists of rolling up the unwanted dimensions into a tiny ball, presumably of the size of the Planck length and time since we want them to play their full part within such an order of space-time.

Such a device is not wholly unfamiliar: our ordinary terrestrial space is, in a certain sense, rolled up on the surface of a sphere; one does not get to infinity by going round the equator. Another familiar image is that of a hose: looked at from a distance it is just a line but close up you can see that you can go round and round its compactified inside dimension. So the story is that our three familiar space dimensions are infinite in extent—begging the general-relativistic possibility of curved and finite space, which is not in question here—as also is the time dimension. But at each point in that familiar four-dimensional world we can also take off in, say, six other unfamiliar dimensions, although if we do, we shall be back where we started, not in no time at all, but in about 10^{-43} seconds.

That is the story, and it can be made quite acceptable in terms of the normal rules of the quantum-mechanical game. But it does leave the great question mark of why six of the original ten dimensions have been compactified in this way and not the other four, which have been left of infinite extent, and why are four dimensions left uncompactified rather than three or five? Superstrings, of themselves, hold no clue. We do not understand this, and until we do, our global comprehension will be imperfect, no matter how well superstrings work in other respects.

Supermembranes?

Having moved away from points to objects of finite extension as the basic entities lying behind the material world, there is no evident reason why we should take a one-dimensional string as the starting point and not move to a fundamental membrane in two dimensions or to higher dimensional fundamental objects since we have learned not to jib at invoking more dimensions than those of which we are directly aware. Such work is in progress, but so far significant advantages of higher dimensional basic entities have not emerged, and they seem to entrain certain significant disadvantages, notably, the probable loss of the natural accommodation of chirality that I mentioned as a pleasing feature of the superstring.

Many Big Bangs?

The third point that I wish to make before turning to our place in our own Universe concerns the range of possible Universes toward whose consideration our present rationalizations have led us. In the simplest sense our discussion in terms of a single, unique Big Bang is somewhat anthropomorphic: why should there have been, whatever its origin, a single Big Bang? Is this assumption no less arrogant than what led, in Ptolemy's time, to an Earth-centerd Universe, or, in Copernicus' time, to a heliocentric Universe? Why should we suppose that we live in a uniquely created Universe? Even the opening verse of Genesis:

Bereshit bara elohim et hashamayim ve'et ha'artez

is usually mistranslated into English as:

In the beginning God created the heaven and the earth

whereas what it really means is, following the New English Bible:

In the beginning of creation, when God made heaven and earth. . . .

The difference between the two versions is very important for us in the present context because the correct translation implies that God has already been around before creating our own Universe, as St. Augustine told us, and may well have created many others previously and may well have created many others since. Let us not arrogate to ourselves a position whose uniqueness cannot be substantiated by any concrete evidence.

One Big Bang—Many Universes

Even within the unwarranted assumption of a unique Big Bang, the scenarios I have presented imply, rather than exclude, that many independent Universes must have arisen. All forms of inflationary scenario, from which, in principle, it seems difficult to escape, lead to the likelihood of an outcome, from a single Big Bang, of many Universes that are either noncommunicating for everlasting, short of the violation of cherished relativistic principles, or may communicate only in the remote future as the passage of time expands their visible horizons' extent to the point of overlapping.

(I might, at this point, remark that there are solutions to the equations of relativity that can be interpreted as admitting, or, some would say, requiring, the existence of superluminary particles, tachyons, that travel at greater than the speed of light and that, some hold, might permit of communication between regions that we normally consider to be causally disconnected. Tachyons are not in vogue at the present time, but we should keep them in the back of our minds.)

Many Forms of Many Universes

There are many forms of multiple Universes. There is the simple inflationary version in which many, perhaps an infinite number, of bubble Universes are generated, existing separately and unconnectedly in what one might loosely call ordinary space and time. Then there are the other inflationary versions, for example, the chaotic inflation of Linde that I mentioned, which invokes spontaneous quantum fluctuations of the scalar field to loose off inflation and may do this infinitely often: a new Universe every time. Then there is Everett's Many Worlds version in which with each quantum act a branching takes place so that there is an uncountable multiplicity of Universes, of which we inhabit only one and define it in terms of our consciousness, and that exist simultaneously and coextensively but that are separate and unconnected in, in the jargon, Hilbert space, which is the abstract space that measures the range of realizable quantum possibilities. Then there is the bouncing Universe that is closed in the sense of having $\Omega > 1$ so that it returns to a Big Crunch from which it emerges, phoenix-like, and sets out again on a limited excursion through space and time but with, having passed through an all-destroying singularity, a completely new set of laws of Nature and natural constants to go with them, and then it continues to bounce again and again. As T. S. Eliot wrote:

> In my beginning is my end. . . .
> In my end is my beginning.

In a sense the bouncing Universe does not require us to ask about the original act of creation because it was always bouncing and always will be; neither does the chaotic inflationary scenario of Linde to which I have just referred.

The Wave Function of the Universe

As I have already mentioned, the concept of the Big Bang singularity is, in its most elementary form, with its zero size and infinite density, a classical concept. It is held by many that the wedding of general relativity and quantum mechanics that we must ultimately achieve will dispose of the singularity in this simple form and that we must look to other scenarios for the beginning and the end that will avoid the singularity because quantum mechanics does not like infinities. Attempts to wed general relativity and quantum mechanics in a quantum gravity have taken many forms, leading to the setting up of a wave function of the Universe such as that following considerations of John Wheeler and Bryce DeWitt. Such an equation has a place in quantum cosmology analogous to that of the Schrödinger equation in microscopic physics.

Boundary Conditions

All considerations about beginnings, whether classical or quantum, involve boundary conditions: given the equations and the laws and the constants how did it actually start off? What was where and moving how fast in what direction when the gun was fired and the equations took over? It is a matter of taste in regard to when we begin to worry about such things, but scientists like to put such questions as far back along the line as possible. It is impossible to disprove P. H. Gosse's contention, in 1857, that the world was created in 4004 B.C. with the fossils already in place in the rocks, but it strikes us as implausible and we prefer theories that start as far back as we know how with as few *ad hoc* conditions attached to them as we can manage.

The simple Big Bang without inflation did pretty well in this respect, but we had to put in a lot by hand; that is to say, we had to make special assumptions about boundary conditions, and as we saw, some of these were exceedingly implausible such as those related to the mean density of the Universe where, to get today's value of Ω we had to fine-tune things in the earliest moments to an incredible degree. Inflation removed that worry but left us with others such as the initial spectrum of density fluctuations on which the generation of galaxies depends.

We should be happiest if we could understand how things got to be the way they are without having to make any special assumptions at all about the boundary conditions at the beginning. The quantum-relativistic wave function of the Universe approach does not, of itself, do this for us but may hold out hope.

Imaginary Time

Thus Stephen Hawking and his colleagues have suggested a formal recasting of the definition of time in which, in the jargon, time is given an imaginary aspect instead of, as usual, a real one. From the mathematical point of view this is no more than a shift in the way in which the symbol we use to represent time is manipulated within the equations; its physical significance is another matter. The shift gets, however, over certain technical difficulties that arise in working out the consequences of the background assumption to the whole approach. This approach is to imagine that in getting from A to B in space and time a particle does not follow a unique path, as it would in classical mechanics, but rather follows all possible paths; the probability of its arriving at B is then given by adding up all the contributions from all the different ways, totting up all the different waves with their different amplitudes. This is, in fact, just the procedure that Richard Feynman originally used in his recasting of quantum mechanics in modern form—the so-called "sum over histories" method.

Many people did not like the sum over histories because it smacked too much of the Everett Many Worlds picture, which was thought to be unphysical, and alternative formulations were devised; however, the sum over histories has subsequently been found very useful in discussing a number of important problems—such as the acceptability of the electroweak unification—and is certainly still with us. But this summing over histories appears to be the only likely candidate for discussing the beginnings of the Universe through a global wave function, which brings the Many Worlds viewpoint back into the limelight.

With this introduction of imaginary time the technical difficulties of the Many Worlds/sum-over-histories approach are smoothed away and the distinction between space and time disappears even formally. In the conventional formulations of classical relativity, for example, we speak of space-time in which space and time are intimately related, and there are rigorously defined trade-offs between them, but they retain their separate characters; this is no longer so if we make time imaginary. The world then consists purely of a four-dimensional Euclidean space just as the domestic world of our household economy consists of a three-dimensional Euclidean space plus the passage of an essentially unrelated time.

Finite Space-Time

In ordinary quantum mechanical calculations about atoms and particles we can use the imaginary time trick and get the usual answers. However, when we use the four-dimensional Euclidean space in our discussion of the early Universe, it opens up the possibility that, because of the equivalence of the four dimensions, space-time might be finite in extent but without a boundary just as the surface of a sphere is finite but has no boundary, no edge. In that case we would not have to specify any boundary conditions because there would not be a boundary on which to specify them. The most likely behavior of the Universe, following this no-boundary-condition assumption, would, in the real time of a real observer, involve a Big Bang followed by something like the inflation that I have already discussed, leading into a closed Universe, with, however, Ω very close to unity, that would eventually, but in the extremely remote future, collapse back into a Big Crunch. In real time there would still be events that would look very much like the singularities that we have already considered, and at which the laws of Nature come into being and have their end, but these singularities would not exist in imaginary time. The meaning of much of this is obscure, but it may point the way to an understanding of the Universe from which arbitrary assumptions are removed.

Quantum Wormholes

Much work on quantum gravity is now cast in the language of this Euclidean four-dimensional space-time in which time is imaginary in the sense that I have just touched upon. I mentioned earlier the wormholes of classical general relativity, which can connect different parts of our Universe but probably play no important role in cosmology. Hawking and Sidney Coleman have argued that in quantum gravity with imaginary time the situation is very different; wormholes are no longer safely tucked away within black holes but become baby Universes that branch off from our own at some point in space and (imaginary) time and then reattach themselves at some other place and time.

Hawking and Coleman have shown that each type of baby Universe entrains a relationship between the interaction of itself with ordinary matter and the cosmological constant, whose size we have seen to be so perplexing. Coleman has argued further that the cosmological constant overwhelmingly the most likely to emerge from this interaction of all possible baby universes with matter is zero. This would very satisfactorily answer my question of where the cosmological constant has gone: It has gone down the wormhole. Superstrings, which I mentioned earlier as perhaps also putting a zero-valued cosmological constant on offer, are not directly involved in Coleman's argument, although they do give some comfort in his use of a nonrenormalizable quantum gravity.

Time

I have just touched upon time and its definition; it is an important, fascinating, and intractable problem. We know, literally from our own experience, that time has a sense: we age, a match flares and dies; we say that that is the direction of the arrow of time—the direction in which things get worse. And yet, as I mentioned earlier in my discussion of quantum mechanics, in the equations of physics on which all natural processes ultimately depend, time does not have a preferred sense and things work just as well with time reversed: in a simple process like the collision of two billiard balls you can run the film backward and see an equally possible collision. But run backward the film of the flaring of the match and that just does not happen, although the microscopic laws of physics would not be violated if it did.

The point is that the flaring of a match is not an elementary phenomenon in which the constituents in the final state are the same as in the original state and simply in different states of motion relative to each other, as in the billiard balls analogy, but is a complex phenomenon involving a very large number of constituents whose state of organization, whose degree of ordering, is very different in the initial and in the final states: the ordering of the atoms in the wood and in the brimstone of the match is patently greater than in the cloud of smoke at the end. Now there are few ways in which order can be achieved compared with the number in which disorder can be achieved, and so it is not surprising that complex systems faced with the choice of spontaneously ordering themselves or spontaneously disordering themselves choose the latter because there are so many more ways of doing it. Of course, if they had to choose *one particular disordered arrangement* out of the very large number of alternatives, that would also be unlikely to be achieved, but that is precisely the point—the flaring match does not, ahead of time, announce the precise specification and structure of the cloud of smoke into which it is going to transform itself; it just goes up in smoke.

There is, therefore, a kind of semilaw of Nature that says that things tend to move in a way in which the degree of disorder is increased; it is only a semilaw because it is not absolute but only statistical: there is a finite probability that the cloud of smoke *will* reconstitute itself into the unburned match, but the probability of this happening is negligibly small; it would not, however, be a miracle because it would not involve a suspension of the laws of Nature but only an exceedingly unlikely happening within those laws. But we should certainly regard it as a miracle and doubtless quote Eugen Rosenstock-Huessy:

What is a miracle? The natural law of a unique event.

The Order of the Universe

The Universe as a whole must, then, be moving toward a state of greater disorder, be running down, until everything is just a cloud of smoke. This is the so-called Heat Death of the Universe that so exercised philosophers a few decades ago before more high-tech forms of Armageddon such as the Big Crunch, black holes, and proton decay had come on the market. The semilaw does not say that order cannot be increased locally within a system, but if that happens, it must be paid for by a compensatory greater increase in disorder elsewhere within the system. That is what is happening when plants and human beings grow, but in the end they all go up in smoke.

The fact that the Universe shows considerable signs of order—galaxies and stars and you and me—shows that it is evolving toward a state of greater disorder from a state of greater order and so must have emerged from the Big Bang in a state of less than the greatest disorder. Now this obviously has to be either arranged or be an accident or follow naturally from the nature of the Big Bang itself. This is another of those boundary-value problems with which all models of the Big Bang are plagued except possibly no-boundary-condition models such as that of Hawking to which I have referred, and that particular one indeed entrains that things started off in a smooth and ordered state, which is an important plus for it.

The Cost of a Universe

A natural question, on contemplating the possibility of a very large number, perhaps an infinity, of Universes, is "how is it all paid for?" The Universes each contain large amounts of matter and large amounts of radiation of various forms and that costs energy. But the matter and the mass equivalence of the radiation acts gravitationally upon itself, and that corresponds to a negative energy, which would be released if there were to be a gravitational collapse. It is a fairly general result from general relativity that these two sources of energy, positive from the matter and radiation and negative from the gravitational interaction, might exactly cancel one another so that the net cost of a Universe is zero. As Alan Guth has put it, it is the ultimate free lunch.

A Philosophical Conundrum

At this point we encounter a significant philosophical conundrum. If we are led, either by our modesty or by our physics, to consider the existence of a large number, perhaps an infinity, of other Universes, with which we cannot communicate because we are cut off from them either in space or in time or both, and of which we can gain no knowledge, should we continue to take them into account in our discussions of the totality of possible creation or should we discount them entirely on the grounds of their inaccessibility and therefore of their irrelevance?

There are those who take the latter line very strongly, flourishing the last proposition of Wittgenstein's *Tractatus*. My own view is that there would indeed be little point in contemplating a multiplicity of causally disconnected Universes if we thought that they were all the same in the sense of operating under identical physical laws with identical sets of natural constants and differing only in the particular paths that their separate evolutions had happened to follow. But that is not the position to which physics leads us.

Universes of Different Nature

Our present understanding, which has not been thought up for this cosmic purpose but which has emerged naturally from our discussions of physics on the laboratory scale, leads us to the conclusion that when the initial central symmetry of the Big Bang itself breaks and when the lower, successor symmetries break, the very Natures of the worlds that are thereby generated severally depend upon the sense in which those symmetry breakings happen to have taken place—remember the arbitrariness associated with spontaneous symmetry breaking that I have stressed several times. We cannot prescribe from a particular sense of symmetry breaking what the particular resultant set of laws of Nature and their associated physical constants will be, but it is certain that the separate disconnected Universes born of a single Big Bang must be different in most of the essential respects that guide their subsequent evolution: different laws operating on different ingredients in different ways and with various numbers of compactified and uncompactified dimensions; all these things are consequences of the ways in which symmetries have happened to break; I repeat: have *happened* to break.

With a very large number of Universes some will be quite like, but not exactly the same as, ours with similar laws and closely similar constants determining their operation, but others will be radically different with laws and behavior that we do not even know how to envision. With an infinite number of Universes there will be an infinite number identical with our own with a book identical to this one being read at this identical page by someone identical to you at this identical moment—although I hope that you will reprimand me for speaking of "this identical moment" when that expression has no sense when applied to causally disconnected systems. But what I want to stress is not the possibility that other Universes may be very similar to our own but rather that the vast majority of them will be very different from ours, most of them unimaginably and incomprehensibly so.

Physics and Metaphysics

Physicists have led themselves into an arena more commonly thought of as that of the philosopher—and the metaphysician at that. But they must not be afraid of it; they have been led there by their understanding of the firmly based natural world of their own experience and have not cooked it all up *ab initio* on the basis of ideas that are not themselves part of a verifiable natural system. As Andrei Linde has written:

Thus the development of physics reveals problems which traditionally were beyond the scope of physics. It seems that to go further we must investigate these problems without prejudice, rather than wait until philosophers try to do it for us.

At one time logical positivism was a popular philosophical standpoint. Metaphysics, defined as propositions unverifiable by observation, was regarded as meaningless; the business of philosophy was restricted to establishing the logical foundations of scientific and mathematical intercourse. But logical positivism, at least in this simple form, has now been abandoned. Even its last strong advocate, Freddy Ayer, concluded that attempts to set up verification procedures that will select and reject metaphysical propositions while admitting those that might respectably be incorporated into the canon of our understanding of the "real world" will not succeed; in other words it is now conceded to be impossible to separate facts, broadly defined, from matters coming through what might be called aesthetic, religious, or ethical channels. Metaphysics, which the physicist, through his very physics, is now taking on board, is back in. But in recognizing this, and, as I do, in welcoming it, let us not forget H. L. Mencken's dash of cold water:

Metaphysics is almost always an attempt to prove the incredible by an appeal to the unintelligible.

The cosmos knows no bottom line.

4. The Goodness of Fit

I now turn to my reason for stressing the vastness of the spectrum of Universes, parallel to, but radically different from, our own, that spreads itself before us in our mind's eye.

The Anthropic Principle

There is an astounding goodness of fit between us and our own Universe. If, as I shall relate, the Universe had been only slightly different in any of many ways to do with the laws and constants of Nature and to do with the properties of the substances to which those laws give rise, we would not be here to wonder at it; our existence seems to be due to the delicate interplay of a large number of individually incredible accidents. This bundle of considerations goes under the general heading of the anthropic principle, which has been developed in a multiplicity of forms by Brandon Carter, Robert Dicke, and many others. I will not attempt to survey this ramification of forms but just state the principle in the crudest possible way: our Universe has to be as it is because if it were not, then we would not be here.

Teleology

There is, of course, a simple teleological explanation of the anthropic principle, namely, that God deliberately created the Universe in such a way that we might come into being in it and give God glory. This is the view reflected in the prayer offered by the Scots Presbyterian minister, a dour breed not above bargaining with the Lord, and recorded in the little inspirational volume *Prayers and Graces:*

Give us the power, O Lord, for if Thou dost not give us the power we will not give Thee the glory, and who will be the gainer by that, O Lord?

But for us to consider that God is in need of glory is surely as arrogant as to think that God must have made do with a single Big Bang. Rather, the concatenation of unbelievable coincidences on which we would depend in a unique Universe might rather lead one to suppose that we are indeed an accident and were not in mind when the Universe was constructed as the last refuge of a jealous God.

The Catalog of Fit

Now to the catalog of our fit to our Universe:

I first of all discuss what kind of kit of chemical parts is needed for intelligent life to come into being and the sort of environment that must be provided; then I discuss the kind of physical Universe that is necessary to provide that kit of parts.

Chemistry. To begin with, a word about the biochemical basis of life. Whatever definition one cares to adopt to categorize life, and this is no trivial problem, it must be clear that intelligent life, even of a fairly lowly order, must involve complex systems. There cannot be any possibility of intelligent life being based other than on a rich variety of chemical structures of the same general type as those with which we are familiar, although the life forms resulting from different varieties of chemical structures might possibly be very different from those we know. Although it is imaginable that synthetic life forms might ultimately be produced in the laboratory, based on different sets of chemical elements, there is general agreement that carbon is the only basis for the *spontaneous* generation of life and that hydrogen, oxygen, and nitrogen are also essential. At all events we must have atoms and molecules of an appropriate variety and flexible nature.

Water. In addition to a carbon base for life it seems that water is essential—with ammonia as an alternative too remote to be contemplated seriously and with no other candidate in sight. In addition to its general role as an enabling agent in almost all the biochemical cookery involved in the putting together of the molecular bases of life, water is probably uniquely suited to such essential activities as the formation of cell walls. But water is a very funny substance in that almost all of its properties, specific heat, surface tension, and so on, are anomalous, either the biggest known or the smallest known. And furthermore, these unusual characteristics are necessary for it to do its biochemical jobs.

Another extremely unusual property of water is that it swells when it freezes: ice floats. This property alone may appear to be responsible for the emergence of life on this, or any similar, planet because if ice did not float but went down to the bottom, such as would be the case on the freezing of almost all other known liquids, the oceans would set solid since the ice would not be exposed to the warmer air when that came around. There would be no help in moving the planet closer to its star so that the temperature did not fall below freezing because then the surface of the planet would be too hot to support life.

Water is a substance of intense academic interest in its own right, and a full understanding of its remarkable properties is not yet available, but it is clear that they depend upon a delicate interplay of the chemical bonding forces, and it is clear that this interplay must have been very finely tuned to endow water with its life-enabling characteristics.

Planets. When we turn to the environment on the larger scale of the planet Earth and of its star we may seem to approach the trivial because there are so many stars in our visible Universe, about 10^{23} from about 10^{12} galaxies at about 10^{11} stars per galaxy, and the likelihood of the formation of planetary systems around stars is probably not, following recent observational hints, as remote as it used to be thought. But it is still useful to pause on this matter if we wish to gain some feel for the overall probability that life might have emerged within our Universe, even granting the suitability of the fundamental starting conditions, namely, the provision of the kit of parts, that I shall shortly address. It is not just a question of a suitably sized planet at an appropriate distance from a suitable star: the planet has to be made of such materials, and in such a way, that it constitutes a congenial setting for the emergence and the evolutionary development of life. Consider just one aspect of this, namely, the atmosphere.

Oxygen is needed for life-as-we-understand-it. But not too much or we oxidize and the planet, literally, goes up in flames: today the oxygen content of the atmosphere is about 21% and slowly rising; already fires, due chiefly to lightning strikes, are a significant factor in the balance between the growth of vegetation on the planet and its removal; if the oxygen proportion were to rise only to 25%, which it will do within the next few hundred thousand years, then fires would win out and the Earth would be denuded of vegetation —and us. In any case, the Earth's atmosphere is only marginally stable and we are perilously poised between glaciation and the greenhouse effect.

Too small a planet will not retain an atmosphere and too large a planet would have so strong a gravitational force that it would break the bones of any creature large enough to carry a brain of adequate size for intelligence to be developed.

Time Scales and Evolution. Given an appropriately sited planet of appropriate characteristics, we must still ask what are the chances of intelligent life emerging in the time available: a few thousand million years, before the star finishes its steady energy generation and changes its characteristics in a way that makes it inimical to life such as shrinking to a white dwarf or swelling to a red giant. There is fairly general agreement among students of evolution that the time necessary for the emergence of intelligent life is of the order of thousands of millions of years so that the parent star in question must remain reasonably constant in its properties over such a period and the planet must also not change too much. I shall return to these considerations.

However, the mere provision of suitable conditions over a sufficiently long period is by no means enough to ensure that intelligent life will emerge. The consensus view puts this probability at so low a level that the chances of such emergence having taken place anywhere else within our visible Universe are small. I believe that this is not just a consequence of our wishing to arrogate to ourselves a unique position in the order of things, as we have done so many times in the past; rather, it is a genuine indication of our remarkable improbability.

Improbable as we are, even when given appropriate conditions for our emergence, those conditions, as determined by the laws of Nature and the numerical values of the constants that those laws involve, must indeed be appropriate and to that I now turn.

The Rule of Law

I have already, in my earlier discussion of the laws of Nature, touched upon the most basic consideration of all concerning our Universe, namely, that it is ruled by law. As I mentioned, what we call the laws of Nature are only our abstraction from, and codification of, what we recognize as de facto orderly and reproducible behavior within the Universe. The Universe appears to be an orderly place in the sense of operating under definable laws that, in particular, represent the reproducibility of Nature's behavior if the same initial conditions are themselves reproduced. There is no obvious reason why this should be so; the Universe might have been essentially disorderly and therefore incapable of organization into the systematic structures, of which we are a well-developed example, or even of manifesting the entities that such structures comprise such as self-consistent particles and radiation. One would then not even have been able to recognize chaos because chaos itself represents only the extreme of orderly behavior and cannot be discussed in the absence of the concept of order. No order, no law, no atoms, no life; that would not do.

Dimensions. The next most basic consideration, granted the rule of law, is that of the dimensionality of our Universe. We inhabit a world of three space dimensions and one of time; more cautiously, we inhabit a world that, as we see and inhabit it, operates to all intents and purposes, so far as its everyday functioning is concerned, in terms of those four dimensions, although we recognize that there may be further dimensions that, as I mentioned when I touched upon superstrings, have been rolled up out of our sight and need for practical concern. We have absolutely no a priori idea of why we have this three-plus-one dimensional world. What if it were not?

More than one independent dimension of time would obviously land us in an irrational acausal world without logic, order, or the possibility of consciousness; that would not do. The catastrophic consequences of more than three dimensions of space, or less than three, is trickier to establish. It is, however, the case that both for the classical gravitation of Newton and for the general relativistic gravitation of Einstein, no stable planetary orbits are possible in more than three spatial dimensions; in three dimensions local gravitational structures such as our solar system are effectively stable (i.e., *pace* the strictures of deterministic chaos) and independent of the behavior of the Universe in regions remote from them, but in more than three space dimensions this is not so and small changes in remote parts of the Universe would throw our local planets off their orbits and into the Sun or into outer space. That would not do.

When we turn from motions in the large to atomic structure, a similar result emerges: stable bound atomic orbits do not exist for more than three spatial dimensions; in more than three dimensions there would be no stable atoms, no chemistry, no life. That would not do.

Could we make do with only two spatial dimensions, as explored by Edwin Abbott Abbott in his sexist sociological extravaganza *Flatland*? Brains are difficult to design in only two dimensions since wires could only intersect and not pass over each other, but information-processing networks might nevertheless be possible. More formidable problems arise when we consider the fidelity of the transmission of information. It is a general result, arising from consideration of the behavior of wave equations in varying numbers of dimensions, that only in three dimensions can sharp signals be transmitted without distortion. For example, in two dimensions (in fact in any even number of dimensions) signals transmitted at different times can arrive simultaneously. In effect, the transmission of information would become acausal. That would not do.

Life in only one spatial dimension need not seriously detain us. It would certainly not do.

We depend, therefore, on there being just the three effective dimensions of space and one of time, but we have no idea why, in any other sense, the world should be made that way, particularly in view of the apparent necessity of higher dimensional bases for a description of a world that was unified in its beginning and of our present inability to understand why the breaking of that unification should have effectively disposed of all but just those dimensions that permit us to enjoy our existence.

I turn from this first mystery of life to more specific considerations of what we are made of and how catastrophically different it could all so easily have been.

The Neutron-Proton Mass Relationship. For life we need complex atoms, which means that neutrons and protons must be stable against their decay into other sorts of particle and must stick together to form complex nuclei around which electrons can circle to make neutral atoms. Now neutrons and protons are particles that are distinguished chiefly by the fact that the one species is electrically charged while the other is not. The strong or nuclear forces that they exert upon one another are almost independent of which particle is which. One manifestation of this equality is that the masses of neutron and proton are the same to within a tenth of a percent. This is why, when the dust of the Big Bang has settled, after a few tens of microseconds, and we are left only with neutrons and protons, they are expected to be present in accurately equal abundances. What happens then I have already described: the neutrons and protons interact with the electrons and neutrinos as the Universe cools, and because the neutron is slightly heavier than the proton, neutrons tend, on balance, to get converted into protons, and we finish up, when this process is frozen out by the fall of the temperature, with a proton-to-neutron ratio of about 7 to 1, following which the light nuclei form.

This happens only if neutrons are heavier than protons. Why should they be? Off hand we might have expected it to be the other way round because if the particles are the same apart from their electrical charges, then the electrostatic energy due to the charge on the proton, converted into mass by $E = mc^2$, should have made the proton heavier than the neutron, which has no electrical charge and therefore no electrostatic energy. This argument is excessively naive and ignores the fact that although the neutron is electrically neutral overall, it, like the proton, contains electrically charged quarks as its constituents, but the argument at least shows that the neutron-proton mass difference is not something that must obviously have the sign that it in fact has.

What if the neutron-proton mass difference had, indeed, been the other way round, which we might more naturally have expected? To begin with, things would not have been all that different; we should simply have emerged from the neutron-proton interconversion phase of things with an excess of neutrons over protons rather than the other way round and the initial building up of the light nuclei would not have been qualitatively changed. But we should have emerged after those first few minutes into a world consisting almost entirely of ^{4}He and stable neutrons rather than ^{4}He and protons. Galaxies would have formed and stars would have shone, but there would have been no free hydrogen possible, and therefore, no water and no life-generating molecules involving hydrogen because such protons as might have been generated would have decayed radioactively into neutrons plus positive electrons if the mass difference were sufficiently great or, if not, would have captured negative electrons to turn themselves into neutrons that way. No hydrogen, no life. That would not do.

The Weak Interaction. Another consideration from the epoch prior to the building up of the lightest elements in the Big Bang concerns the strength of the weak interaction that, through the mediation of electrons and neutrinos, converts neutrons into protons during the first few seconds. As we have seen, the strength of the weak interaction balances rather nicely the rate of cooling of the expanding Universe so that we come out of this interconversion phase with comparable numbers of both neutrons and protons, quite a lot of neutrons, although heavier, remaining: 7 protons to 1 neutron. If the weak interaction had been considerably stronger, then the interconversion reactions could have held their own for longer against the falling temperature, and so we should have emerged with almost 100% protons. That probably would not have mattered very much because the ^{4}He, made primordially in the Big Bang, has no special function in the subsequent development of the Universe so far as we know.

If, however, the weak interaction had been considerably weaker, it would have been a very different story because then we would have emerged into the epoch of light element synthesis with almost none of the neutrons converted into protons so that the world would have started out toward galaxy formation consisting of almost 100% ^{4}He. No free protons, no hydrogen, no life. That would not do.

We can only answer the question of why the weak interaction has the effective strength it has in the relevant low-energy regime of interest somewhat rhetorically, within the framework of the electroweak unification, by reference to the masses of the W and Z particles, and since we have no idea why those particles have those masses rather than much greater ones, we cannot say that we understand why the weak interaction is not much weaker than it is.

Neutrino Species. Another consideration, to similar effect, concerns the rate at which the Universe cools. As I have already remarked, this depends upon the number of degrees of freedom offered by the various species of particle available for the sharing out of the thermal energy that the temperature represents; the more the species of particle, the more rapidly the temperature falls. But if the temperature falls very rapidly we shall finish up with almost equal numbers of neutrons and protons, just as we have seen we should if the strength of the weak interaction were significantly less than it is, and, as in that case, this would result in a Universe useless for the emergence of life, consisting just of ^{4}He.

Now among the species of particle contributing to the number of degrees of freedom are the neutrinos, of which we know three, namely, those associated with the electron and with the heavier electron-like particles, the muon and tauon respectively. As we saw earlier, these three species enable us to understand very nicely the cosmic abundance of ^{4}He, although one more species might, at an uncomfortable squeeze, be admissible. But we have no idea why there should be just three or, possibly but improbably, four species of neutrino rather than any other number. If there were more than about eight such species, the Universe would have cooled so rapidly that, even with the value of the strength of the weak interaction as it in fact is, we should have been left with almost equal numbers of neutrons and protons and then a Universe of ^{4}He. That would not do.

The Nuclear Force. When we come to consider the strength of the nuclear force in relation to the building up of the light elements, both in the Big Bang and in the stars, we encounter a fantastic piece of fine tuning. For this building up to take place it is necessary for the deuteron to be stable against breakup into a neutron and a proton because the deuteron is a unique step on the way to the building up of the heavier elements both in the Big Bang and, more importantly, in the stars: no stable deuteron, no heavier elements, no nuclear energy generation in the stars. That would not do.

Now the deuteron, although energetically stable, is only just so: it is bound by only about 0.1% of its mass energy; if the nuclear force had been only very slightly weaker—by a very few percent—than in fact it is, then the deuteron would indeed have been unstable. Now consider the state of affairs if the nuclear force had been stronger than in fact it is. If the strength were increased only very slightly—again by a very few percent—then not only would the deuteron have been stable but also two protons would have bound stably together to make ^{2}He. All the protons would have combined into ^{2}He during the first minutes after the Big Bang and no hydrogen would be available today. That would not do.

So our existence depends upon the fact that the strength of the nuclear force falls into that tiny window, a very few percent wide, out of an infinity of choice, that makes the deuteron stable but does not bind two protons together. We have not the faintest idea why this should be so.

Complex Nuclei. The stability of complex nuclei, and therefore, the existence of the heavier elements, depends upon the interplay between the strengths of the electrical and nuclear forces, both of which have had to be taken into account in the remarks I have just made about the stability of ^{2}He. (Without the electrical repulsion between the two protons it would have required an increase in the nuclear force of only less than 1% to make ^{2}He stable.) This interplay brings the electrical force more and more into prominence as the nuclei get heavier because it is of long range so that every proton in a nucleus interacts with every other one, and the electrical energy therefore increases as the square of the number of protons in the nucleus, whereas the nuclear force is of short range so that a neutron or proton interacts only with its nearest neighbors, and the nuclear energy therefore increases only linearly with the number of particles.

It is this that brings about an end to the periodic table of the elements: when nuclei are so highly charged that the disruptive effects of the electrostatic repulsion overcome the binding effects of the attractive nuclear force. If the electric charge were to be increased by only a factor of three (but why should it not be? or by a factor of a hundred), then the periodic table would consist only of hydrogen, helium, lithium, beryllium, and boron: no carbon, no life. That would not do.

Electrical Charge. I have now mentioned the electrical force: we have no idea of why it takes its particular value as parameterized by the charge on the electron. If it had been larger by that small factor of three, we should not be here. But what is so astonishing that it is almost never mentioned is the equality of the electrical charge in its various manifestations. The electrical charge is, in fact, equal in magnitude on the electron and on the proton to within one part in 10^{21}. This cannot be understood except in the context of GUTs, which permit the interconversion of quarks and leptons so that if charge is to be conserved (but why should it be?), there must be exact equality of charge between protons and electrons.

It is very important that the Universe should be electrically neutral overall because of the enormous difference in strength of the electrical and gravitational forces. As a measure of this: if there were suddenly to be brought about an imbalance of only part in 10^{18} between the electrical charges of the electron and the proton (which would not seriously upset atomic and molecular structures) the resultant overall electrical repulsion between a person and the Earth would just cancel the gravitational attraction and cause us to drift off gently into space.

An unimaginably tiny net overall electrical charge to the Universe would dominate its development and structure. We have no idea why it should not have such a net charge except in terms of that kind of consideration of cosmic democracy that asks why it should be one rather than the other.

Making the Carbon. I turn now to more detailed considerations. Carbon is essential for life: how does it get made? Production in the primordial processes of the first few minutes of the Big Bang is negligible and we must look to the stars. The stars generate their energy quite largely by burning hydrogen into helium. Protons combine with other protons through the weak interaction, giving deuterons plus positrons, or by capturing electrons to the same effect. The deuterons then interact rapidly with each other and with other protons through various fast nuclear reactions to yield ^{4}He. But then we are stuck because neither ^{5}He nor ^{5}Li, which might be made by sticking a neutron or a proton onto ^{4}He, is stable, so we have encountered a bottleneck. Nor do we get anywhere by trying to stick two ^{4}He nuclei together because the resultant ^{8}Be is also unstable. However, as Ed Salpeter pointed out in 1951, the fact that ^{8}Be is only just unstable—by about 90 keV —means that its lifetime after formation in the collision between two nuclei of ^{4}He, although finite, is quite long on the nuclear timescale—about 10^{-16} seconds—so that it can be regarded as a short-lived target for another ^{4}He to come along and form the desired ^{12}C.

This already smacks of a remarkable dispensation on the part of Nature to breach the bottleneck at mass number 5 since just a little less tendency toward binding in ^{8}Be would have given so short a lifetime as to be useless. However, even with an ^{8}Be that lasts this long, the reaction rates to be expected for the attachment of another ^{4}He are far too low to be useful for the building up of significant amounts of ^{12}C unless, by a further miraculous dispensation, there happens to be an excited state of ^{12}C so close to the energy represented by adding ^{4}He to ^{8}Be that the reaction will proceed resonantly through that state at the temperature of the interior of stars, the nuclear reaction cross sections being enormously enhanced under such resonant circumstances. But that is not all: not only must the excited state of ^{12}C be present at the right energy, it must also possess the right quantum properties of intrinsic angular momentum, namely, zero, and also the right reflection properties, parity in the jargon, namely, even; otherwise the ^{4}He would not be able to get into the ^{8}Be to make the ^{12}C.

All this was pointed out by Fred Hoyle in 1953 who concluded that to build up significant, and adequate, amounts of carbon in stars, there must exist such an excited state of ^{12}C at about 7.7 MeV. Such a level is, in fact, found, with the correct properties, at 7.65 MeV. It is difficult to overestimate the astonishing nature of this coincidence, which provides the, literally, vitally essential energy level correct to within a few kilovolts of excitation when the nearest levels to it are some 2 MeV above and 3 MeV below respectively; a tiny change in the strength of the nuclear force or of the electric interaction would displace the resonant state into an inaccessible region of the excited-state structure of ^{12}C: no carbon, no life. That would not do.

To practitioners in the field of nuclear structure the affair is even more remarkable because the state in question is unwanted in the sense that it is not predicted at all by the considerations that work very well in describing the low-lying level structures of neighboring nuclei and, indeed, the other low-lying states of ^{12}C itself; it is only by taking higher order effects into account, and that to a rather surprising degree, that one can understand how such a state could be anywhere in that region at all, let alone just where it is needed; it has obviously "been put there."

Keeping the Carbon. Having made the ^{12}C we must keep it and understand why it is not, in its turn, burned into ^{16}O by capturing another ^{4}He just as it was itself made by the capture of ^{4}He by ^{8}Be. In fact ^{16}O does possess an excited state perilously close to the excitation energy at which such consumption of the ^{12}C would take place resonantly, and another slight change in the relevant strengths of the interactions would have resulted in there being no ^{12}C, no life. That would not do.

Distributing the Carbon. From our point of view, however, there is not much point simply in having produced carbon inside stars; it has to be got out and made available for incorporation into planets and into us. This distribution process takes the form of the ultimate explosion of some stars in the supernova process that I sketched earlier. Such supernova explosions are accompanied by the generation of a gigantic number of neutrinos in the core of the supernova as electrons are captured by protons to become neutrons. This was verified by the detection of neutrinos, following the explosion of SN1987a, in apparatuses in Japan and in the United States that had been built to search for the decay of the proton as predicted by the GUT theories as I discussed earlier. It is thought that it is the interaction of these neutrinos with the outer layers of the supernova, conspiring with the outwardly moving shock wave, that blows these outer layers into space and spreads their constituents throughout the galaxy. That is how we come to be stardust. But if the neutrinos have a critical role in all this, now comes the delicate part. If the weak interaction that determines the cross section for the collisions of the neutrinos with the matter through which they are passing is too large, then they will never reach the outer layers of the supernova but be trapped. If, on the other hand, the weak interaction is too weak, then the neutrinos will pass through the outer layers without blowing them away. The carbon would stay in the collapsed star and not get out into you and me. That would not do. Getting it just right involves the nice balance between the weak and the gravitational interaction strengths, together with the masses of the electron and the proton. We have no idea why it should come out just right.

The Fine Structure Constant. An important relationship between the natural constants that permeates the whole of science is the fine structure constant, $e^2/\hbar c$. This is a dimensionless constant and so we can discuss it without any reference to definitions of units and so on; this will be particularly interesting when we come to look at the possible variation of the constants of Nature with time. The numerical value of the fine structure constant is close to $1/137$, and it does not come into the considerations that I am now addressing, namely, the interrelationship of the natural constants, until we reach the realm of Grand Unification on which, although we have no direct knowledge of it, so much has been seen to depend.

Now the Grand Unification symmetry must be broken after the Planck Time; i.e., the mass of the X particle must be less than the Planck Mass; otherwise we should lose our understanding of the way in which the Universe emerged from the Big Bang. This condition requires that the fine structure constant be bigger than 1/180. It is also obviously essential that the lifetime of the proton, against its spontaneous decay that we saw GUTs to entrain, should be greater than the few thousand million years needed for the development of a material Universe appropriate for the emergence of intelligent life in the way that we have already examined. This condition requires that the fine structure constant be less than 1/85. But for that extraordinarily tight "coincidence," $1/85 > 1/137 > 1/180$, of which we have no understanding at all, we should not be here. That would not do.

The Mass of the Electron. We must also examine the electron mass, as well as its charge. The fact that the electron mass is very small relative to that of the proton (about 1/2000) means, crudely speaking, that when an electron goes round and round a nucleus to form an atom, it does not shake the nucleus from side to side very much. This in turn means that atoms can bond together into ordered structures such as molecules and crystalline lattices. If the electron's mass were comparable with that of the proton, this would not be so; molecules would, at best, be wobbly, indeterminate things, and there would be no possibility, for example, of precisely replicating DNA, on which the higher levels of life depend. That would not do.

As I have stressed in introducing supersymmetry and elsewhere, we have no idea where mass comes from—in quantum electrodynamics the electron mass is infinite before the trick of renormalization is played; its de facto mass that we simply insert into our equations when we calculate atomic and molecular structures is a complete mystery, but if it were not small enough we should not be here. That would not do.

The Temperature of the Stars. Another factor on which the emergence of life depends concerns the temperature of stars. Stars are necessary in order to keep us warm, but they are also necessary to bring about the photosynthesis upon which evolution has depended. That, in turn, means that there has to be an appropriate match between the surface temperature of a star, which determines the spectral composition of the light that comes from it, and hence the energy of the photons that it radiates, and the excitation energies of the atoms and molecules whose chemical reactions the star's photons must stimulate for the evolution of life to take place. If the star were too hot, the photons would have too high an energy and would destroy any living organisms: a superultraviolet sterilizer. If the star were too cold, then the photochemical reactions would go too slowly and evolution would not take place at any significant rate. That would not do. The "coincidence" that provides the necessary dispensation involves the fine structure constant, the gravitational constant, and the electron and proton masses; we have no idea why they should combine so congenially.

The Lifetime of the Stars. Another evident necessity is that, as I have already indicated, stars and planets must last long enough for evolution to come about: of the order of thousands of millions of years if we are aiming at intelligence. There is no obvious reason why stars, even without worrying about the GUT decay of the nucleons out of which they are made, should live that long; indeed many do not; stellar lifetimes go roughly as the inverse square of the mass; some stars last only a few million years. But ordinary stars such as our Sun, which are the most abundant type, in fact live for the few thousand million years required for the evolution of intelligence upon a nearby planet if such, by chance, is appropriately available, before exhausting their hydrogen fuel and, if they are light, moving toward a quiescent old age, unable to sustain their planets in the style to which they had become accustomed, or, if they are heavy, letting it all rip in a supernova explosion, which is also obviously not of much use to us if we are living nearby. So there is, in fact, a good match between the evolutionary timescale and a typical stellar lifetime. If stellar lifetimes were less than the evolutionary time scale, that would not do. Stellar lifetimes depend upon a combination of the gravitational and fine structure constants and the particle masses with no reason why they should combine in the way in which they so felicitously do.

I turn now to the Universe on the grander scale and examine some of its all-pervading properties rather than those that concern those more immediately domestic issues of nuclei, atoms, molecules, and stars that we have considered so far.

The Density of the Universe. The first issue is the value of Ω, the mean density of the Universe measured in units of what would just close it, just bring about its eventual gravitational collapse. Before examining this I should remark that we can never be certain about Ω on a purely observational basis because all that we can, by definition, determine is the mean density of the observable Universe, which is growing all the time, so that Ω is a matter for continuous revision: today we may think we are going to expand forever, but tomorrow's new slice of, to us, new Universe (old Universe actually, of course) may suggest that we are going to collapse. But I am not here interested in such formal delicacies and wish to be much cruder.

As we have seen, Ω today lies somewhere in the range 0.1 to 2, and I have already indicated how astonishing it is that it is so close to unity. Inflation provides an explanation, but let us for the moment ignore that and give ourselves the freedom to contemplate the whole vast range of in-principle possibilities for Ω from zero to infinity. Within that range the window that permits our existence is quite narrow: for $\Omega > 10$ the Universe would have collapsed to the Big Crunch before stars could have formed; for $\Omega < 10^{-3}$ the Universe would have expanded so rapidly that matter would not have been able to condense into galaxies. In neither case would that do. And yet a range of a factor of 10^4 out of infinity is really rather a small window through which to have to glimpse our destiny.

The Value of the Cosmological Constant. An issue related, in its effect, to the destiny of the Universe is the value of the cosmological constant Λ, which I have mentioned before. This constant, whose sign is indeterminate, operates somewhat like Ω but accelerates the expansion of the Universe or decelerates it depending on its sign. If Λ were too large in magnitude it would bring about, depending on its sign, either a catastrophic collapse of the Universe or so rapid an expansion that nothing of interest could happen. The largest value that Λ could have for these catastrophes to be avoided would be 10^{117} times less than what I earlier indicated we might a priori have expected it to have. If Λ were within a factor of 10^{117} of its commonsense value that would not do.

The Photon-to-Nucleon Ratio. The remarkably large ratio of the number of microwave background radiation photons to nucleons in the Universe, namely, about 10^9 to 1, I discussed earlier; we saw that it could be understood in terms of the not-quite-equal numbers of quarks and antiquarks produced by the decay of the X and $\bar{X}$ particles shortly after the Big Bang. But we also saw that there is no reason at present quantitatively understood why this number should assume this particular value; indeed the simplest expectation, without the intervention of the asymmetry in the decay of the X and $\bar{X}$ particles into their respective quark and antiquark channels, would have been infinity.

Such a Universe consisting entirely of radiation without neutrons and protons clearly would not do, but neither would a Universe in which the photon-to-matter ratio was not small enough. The point arises from consideration of the respective mass densities of the radiation and the matter. If the mass density of the matter is much greater than that of the radiation, then the Universe evolves chiefly through the gravitational effects of the matter upon itself; galaxies form and stars form within them, starting from small mass fluctuations followed by their gravitational growth and condensation. If, however, the mass density resides chiefly in the radiation, then such matter condensations do not take place; galaxies and stars do not form. That would not do.

Now the balance between the density of matter and the density of radiation changes as the Universe expands: the matter density falls as the inverse of the volume of the Universe because the mass of a nucleon does not change as the Universe expands, but the density of the radiation falls with an additional inverse power of the radius of the Universe because, crudely speaking, each photon is "stretched" by that expansion so that its energy, and therefore its mass, falls inversely as the radius of the Universe. So a Universe that is originally radiation dominated will eventually become matter dominated, and only after that crossover time can galaxies and stars form.

That crossover time must evidently be sufficiently long ago to have given galaxies and stars and planets and ourselves time to have evolved, that is, a few thousand million years. In fact, with the de facto photon-to-nucleon ratio of about 10^9, the crossover time came about 100,000 years after the Big Bang and all is well. But the crossover time goes as the square of the photon-to-nucleon ratio, so if this ratio had been increased to 10^{11} or more, the Universe would still be radiation dominated and we should not be here. As I have earlier remarked, in another context, with a range of infinity to choose from, it is more than strange that we should find ourselves within a factor of 100 of the critical value.

The Formation of Atoms. A somewhat similar consideration arises in relation to the epoch at which the temperature of the Universe had fallen to a low enough value for atoms to form out of nuclei and electrons: at earlier times the Universe would be so hot that the radiation would remove electrons from nuclei as soon as they attached themselves. This also, presumably by coincidence, happens to be at about 100,000 years for the photon-to-nucleon ratio of 10^9, although it depends on a different combination of the natural constants from the radiation/matter crossover time that I have just discussed. It also depends on a different power, namely, the square root, of the photon-to-nucleon ratio, so to prevent atoms having formed to this day, and therefore no molecules or life, the photon-to-nucleon ratio would have had to have been 10^{17} or so. But then, of course, galaxies and stars would not have formed anyway, so we should not have to worry. However, this time after the Big Bang at which atoms can come into being does depend on the natural constants, which, with a photon-to-nucleon ratio of 10^9, might have taken values that would have today given us a matter-dominated Universe but a Universe still without atoms and molecules. That would not do.

Galaxy Formation. The final cosmological consideration upon which your reading of this book depends concerns the sort of inhomogeneities that there must have been in the density distribution of the early Universe in order that galaxies might ultimately have condensed out. For such condensations to have occurred, these early inhomogeneities must have exceeded a certain value or the expansion of the Universe would have gone too far before they could have taken hold to bring about the condensation. That would not do.

If, on the other hand, the inhomogeneities had been too large, then there would have been premature collapse of the matter fluctuations into black holes before stars could form. That would not do either. In fact the balance between dissipation without galaxy formation and collapse into black holes is quite a nice one and the degree of permissible inhomogeneity is a factor of only about 10 either side of the value that is inferred for our de facto Universe. Our presence here depends on that quite small window; we have no idea of why it should have opened.

The Generality of the Coincidence

What should we make of this remarkable catalog of coincidences that measures, in so many independent ways, the tight but, from our point of view, necessary fit between between us and our Universe?

I should emphasize that the fit is indeed between us and the spatial whole of our Universe and not just now but always. That is to say that it cannot be argued that we simply occupy a niche in space and time within the Universe as a whole in which conditions have been propitious for our emergence. This is true in the trivial sense that we live on a suitable planet that happens to be suitably near to a suitable star. But it is not true in the sense that we might also be thought to happen to live in a part of the Universe, and at a particular time, where and when the laws of Nature and their constants have taken the right values to bring about all the necessary coincidences. Many of the coincidences refer to the Universe in the gross, not just to our particular bit of it, and our ability to give a rational account of the whole of the visible Universe within a single set of physical laws and their associated constants shows that those laws and constants cannot change significantly from place to place and from time to time within it.

The Lack of Change

It is useful to pause at this point to emphasize that the laws of Nature are indeed unchanging and that their constants are indeed constant, both statements being made to within tight limits of experimental verification at least within relevant epochs. I make the last qualification because we clearly cannot speak of what the laws might have been like in times to which we can have no access even by inference, which, at the moment, means within the Planck Time or possibly as late as the inflationary epoch. But from, at the latest, 10^{-30} seconds or so we can state with confidence that no major change in the kit of laws of Nature has taken place and, with varying degrees of confidence in relation to various of the laws, that their constants have not changed by more than so much (very little, in fact, as we shall now see) since various epochs.

We have, for example, seen that our understanding of the building up of the light elements following the Big Bang, which covers the period ranging from a few seconds to a few minutes after the starting gun, is based on the fact that the strengths of the weak and nuclear interactions, and the associated cross sections for a whole range of nuclear reactions, were the same then as at our present epoch within which we have measured them and from which we have generated that understanding.

We like to think that things that are right, in all fields of our experience and concern, are unchanging. St. Augustine finds comfort in his ability to address God as

Thou beauty so ancient and, withall, so fresh.

We must, therefore, watch out, as I stressed earlier, that belief in an unchanging order of the physical world is not simply a straitjacket of prejudice into which we insist on cramming natural phenomena.

Do the Constants of Nature Change with Time?

There has been considerable interest in the possibility that the constants of Nature might, in fact, change with time. One of the origins of this interest is the so-called "big number coincidences." Scientific equations involve combinations of physical quantities like masses and strengths of interactions, raised to various powers, plus numerical constants, pure numbers, like π and $\sqrt{2}$ and so on. These pure numbers are usually not very enormous or very tiny. Similarly, other pure numbers that we can construct out of the physical constants themselves are usually neither enormous nor tiny; thus the fine structure constant, $e^2/\hbar c$, is about 1/137, as I have remarked.

However, Nature also contains some gigantic pure numbers. Thus the electrical force between an electron and a proton divided by the gravitational force between them is about 10^{39}. How can we understand such a gigantic number? Similar gigantic numbers crop up elsewhere in totally different contexts. Thus the time since the Big Bang divided by the "embezzlement time" corresponding to the mass of the electron, $\hbar/mc^2$, is about 10^{38}, and the size of the visible Universe divided by the classical radius of the electron, e^2/mc^2, is about 10^{40}. The times, in units of the Planck Time, at which the Universe had cooled to an energy corresponding to the masses of the proton and electron are about 10^{37} and 10^{43}, respectively, which nicely straddle 10^{40}. There seems to be something special about the huge number 10^{40} or so that crops up in so many different ways in connection with fundamental quantities. Also, the number of nucleons in the visible Universe is about 10^{80}, which is just about the square of the previous large numbers.

It occurred to Dirac in 1938 that there must be some deep and invariant connection between such similar huge numbers that arose in such different contexts. But now two of these numbers involve factors that change with time, namely, the time since the Big Bang and the number of nucleons in the visible Universe, so if the relationship to the other numbers, the ratio of the electric and gravitational forces and so on, is itself to be invariant with time, then some or other of the constants of Nature that go into the big numbers must themselves change with time. Since gravity is the force least studied and understood, Dirac plumped for it to change: it would have to vary as the inverse of the age of the Universe to preserve the big-number relationships; that is to say, the strength of gravity would, today, have to be decreasing at the rate of about one part in 10^{10} per year: it is not; astronomical observations on the orbital periods of the Moon and of Mars and on the binary pulsar PSR1913+16 show that the rate of change is, at most, five times less than this.

Much closer limits can be put on possible changes with time of other constants or combinations of constants. It has been clear since the middle 1950s that the major constants that determine the impact of the electromagnetic, weak, and strong forces cannot be changing by more than one part in 10^{10} per year or so. However, much tighter limits can be placed by analysis of the debris of the natural nuclear reactor at the Oklo uranium mine in Gabon, as was first shown by I. Shlykhter in 1976. In this area the uranium deposit is of unusually high concentration, so much so that in the remote, but geologically accessible, past, about two thousand million years ago, the concentration of the thermally fissile isotope ^{235}U (which has a half-life of about 10^9 years) was so high that the ore body functioned as a natural thermal nuclear reactor for millions of years, as is evident from the fission product remains. The important point is that in the region of the reactor the abundance ratio among the samarium isotopes is very significantly different from that found in normal deposits; in particular, the ratio of ^{149}Sm to ^{147}Sm is some ten times lower than in normal samarium. This is because ^{149}Sm has an enormous cross section for the capture of thermal neutrons, which largely destroyed it during the operation of the reactor.

This large cross section is due to there being a strong resonance at a neutron energy of about 0.1 eV. The fact that 2×10^9 years ago this resonance was at closely the same energy as today, as it must have been to bring about the observed depletion of ^{149}Sm at Oklo (the data do not permit of its having shifted by more than 0.05 eV at most), shows that the forces that determine the structure of the nucleus cannot themselves have sensibly changed over that period. Now a typical measure of the strength of the nuclear force is 100 MeV, so the limit placed by these data on its change is something like $0.05/10^8$ in 2×10^9 years, that is, a very few parts in 10^{19} per year. And since the relative contribution of electromagnetic energy to the structure of the samarium nucleus is about 5%, it is also evident that the electric force cannot have changed by more than about one part in 10^{17} per year. Even though the contribution of the weak force to nuclear structure is relatively so small, it is finite, and we may similarly limit its change to about one part in 10^{12} per year. These results show that the constants of Nature cannot have changed significantly since the Big Bang.

Although they are beyond our ken, John Barrow has considered the compactification of the extra dimensions that arise in the context of superstrings and has concluded that the limits on the change of the natural constants that I have just discussed preclude a change in the sizes of those compactifications of more than about one part in 10^{19} per year. It is nice to know that we can rely so much on something of which we can have no direct knowledge.

Conspiracy?

We must always recognize the possibility of conspiracy in these things whereby the natural constants change individually in such a way that their relevant combinations remain accurately constant, but this is not an appealing hypothesis and can be largely ruled out by analyzing phenomena that depend on different such combinations. Despite the fact that the only truly meaningful changes would be those occurring in dimensionless numbers, such as the fine structure constant, we leave this study with the powerful feeling that the constants are accurately constant, at least within the time scale of the age of the Universe, and that is all that matters if we are concerned with the coincidences upon which our existence depends.

How to Understand the Goodness of Fit

We have to accept that our Universe, as a whole, is just right for us in a large number of independent and tightly constrained ways and that our place within it is special only with respect to our suitable planet and suitable star.

I have already mentioned the obvious teleological explanation of final cause, that is to say, that in the beginning of the Universe God made the laws and the constants as they are so that we might be here. As Fred Hoyle has said:

The Universe is a put-up job.

There have been endless and largely sterile disputations through the centuries about God's role, if any, in the development of the Universe after starting it all going: can God, should God, does God intervene from time to time in what would be, by definition, a suspension of the laws of Nature, in common parlance a miracle? One should not greet with mere condescension the child's question "Can God make a stone so heavy that he cannot lift it?"; so many subtle statements of God's material power reduce to that same question under analysis. If miracles could be analyzed they would not be miracles. For me, the natural world is so wonderful that I can only echo Walt Whitman:

I know of nothing else but miracles.

I earlier insisted upon the importance of keeping an open mind about the possible relevance of nonnumerical influences in our world and the role of feeling in our interaction with the Universe; only the most extreme reductionists, of whom there are many, can keep God out of our feelings. This is the point at which I should recall the importance of feeling and reinforce my insistence upon it.

The other way to understand our fit to our Universe is in terms of the various forms of multiplicity of Universe that we have recognized. As we have seen, even a single Big Bang may well, through some inflationary mechanism or otherwise, spawn an infinity of Universes, each with different sets of laws and constants and numbers of dimensions. We simply exist in one of that infinity of Universes in which everything has just fallen out right for us. We have also seen that a single Universe might bounce through an infinity of forms to give at last the one that is suitable for us and that will, at another bounce, be gone. Linde's chaotic inflation model gives rise to an infinite number of differing Universes without beginning and without end, although each individual Universe has its beginning and may, or may not, have an end; again, here we are in one that suits us.

The Universes of God

The problems of taking the discussion any further are certainly metaphysical or theological and bring us back to the role of God. If we limit our consideration of God to God's role in the kind of physical Universe with which we have been chiefly concerned— the physics God (not necessarily the physicist's God)—ignore miracles and any intervention in the laws of Nature, and say that God's business was to start it all off with the right laws and the right constants, then our deepening understanding certainly seems to be confining God to earlier than the time of about 10^{-10} seconds, at which the electroweak symmetry broke, or to the 10^{-35} seconds, at which the GUT symmetry broke, or to the Planck Time of 10^{-43} seconds, at which the ur-symmetry of gravity-plus-the-rest, possibly something like the superstring symmetry, broke, because it was at those times that the laws and their constants were established; quantum gravity or a theory of everything will permit even the sanctum of the Planck Time to be penetrated.

That would still leave the Big Bang itself, the initial act of creation, which, in physicist's parlance, is the setting of the boundary conditions, as the act of God. It is interesting that at a conference on cosmology organized in the Vatican in 1981 Papal approval was given to the idea that God might be limited (the physics God, that is) to just this role of the setting of the boundary conditions. John Paul II said:

. . . any scientific hypothesis on the origin of the world, such as that of a primeval atom from which the whole of the physical world derived, leaves open the problem concerning the beginning of the Universe. Science cannot by itself resolve such a question: what is needed is that human knowledge that rises above physics and astrophysics and which is called metaphysics; it needs above all the knowledge that comes from the revelation of God.

In his speech John Paul quoted Pius XII saying, in 1951, apropos the beginning of the Universe:

We would wait in vain for a reply from the natural sciences who, on the contrary, admit that they are honestly faced with an insoluble enigma.

It is therefore particularly interesting that it was at this same Vatican conference in 1981 that Stephen Hawking introduced his idea, which I discussed briefly, that one might, through the use of imaginary time, set up a scenario in which there were no boundary conditions, no beginning to the Universe.

When, as I have been doing in this book, the physicist considers the beginnings of things and our presence within it all, it is natural to contemplate God. As Francis Bacon says:

A little philosophy inclineth Man's mind to atheism; but depth in philosophy bringeth men's minds about to religion.

But it is not easy to say what a physicist, as a physicist, means by God; when one mentions God one usually means "I am not going to consider these matters in this issue of the *Physical Review.*" Beyond that, and in what I consider to be the assumption of a serious responsibility, the physicist may well follow St. Augustine and say that God is not to be discussed in terms of space and time and that we should not strive to find a location for God within our normal coordinate systems, neither a specific location nor, indeed, one of immanence.

Or we may wish to see God as integral with our physical Universe, representing and constituting its structure and laws, guiding and nudging. But then we must contemplate the possibility, very real in my view, of an infinity of Universes, operating through an infinite spectrum of different laws, constants, and structures, and ask how there can then be an infinity of different Gods each matched to the particular Universe of which the God is a part and that the God controls. But perhaps, as seems very reasonable, when the different Universes arise through the breaking of the symmetries in different ways or their emergence from the singularities in different ways, there are certain common elements, certain things that are the same no matter how the symmetries break and so on, then perhaps God is within those common elements, the same for all Universes. Or, on a different tack, we may invoke, as I have perhaps tended to do by implication, that shadowy nonnumerical component to our experience and to the functioning of natural phenomena, which I have tended to equate with feeling, and say that this shadowy place is God's residence, within our Universe but beyond our equations and mathematics.

I would be inclined to leave it with Ralph Waldo Emerson:

God is our name for the last generalization to which we can arrive.

But we must hearken to Job's stern admonition:

Canst thou by searching find out God?

In this book I have attempted to demonstrate the intimacy of the relationship between ourselves and the world; I must leave open the question of whether the relationship between God and the world is that seen by Karl Marx in the world picture of Lucretius:

Nature without gods, gods without a world.

Taking Stock

It is time for me to take stock. I have called this book "Our Universes" for what I hope are by now obvious reasons. "Universes" is in the plural but we are in the singular. We have only ourselves through which to gain access to those Universes. We have seen over and over again how our nature must impose itself upon, and express itself through, the tale we tell about how the world is made. Over and over again my account of how we tell that tale has used those very human words: judgment, revulsion, prejudice, obsession, taste, beauty, fashion. Our view of our Universes can only express our beliefs, but beliefs cannot be objectified. So do we gain access to those Universes or do we, with great humility, construct them? I think that we construct; I think that they are, indeed, our Universes: the Universe as artifact.

Within philosophy itself, and without benefit of knowledge of the Higgs field, the Planck Time, spontaneous symmetry breaking, and inflation, parallel views are arising. As Jerome Bruner writes:

(The central thesis of constructivism) is that contrary to common sense there is no unique "real world" that pre-exists and is independent of human mental activity and human symbolic language: that what we call the real world is a product of some mind whose symbolic procedures *construct* the world.

And as he has also written:

My own research has taken me . . . into the strategies by which ordinary people penetrate to the logical structure of the regularities they encounter in a world that they create through the very exercise of mind that they use for exploring it.

I think that Henri Poincaré got it right:

A reality completely independent of the mind which conceives it, sees or feels it, is an impossibility. A world as exterior as that, even if it existed, would for us be forever inaccessible.

In other words, reality is a metaphysical abstraction.
I will let William Cowper sum up for the poets:

Knowledge is proud that he has learned so much;
Wisdom is humble that he knows no more.

and I. I. Rabi for the scientists:

Many lessons can be drawn from the evolution of scientific thought and knowledge of nature. The first of all is the lesson of *humility*.

Postscript

One day, toward the end of his life, Bertrand Russell got into a London taxicab. The driver, recognizing the great sage, said "Well, what's it all about then?" For the first time in his life the great sage was speechless.

Index